T0252879

COKE OF NORFOLK
(1754–1842)
A BIOGRAPHY

Thomas William Coke ('Coke of Norfolk') (1754–1842) is best known as one of the main promoters of the improved farming of the 'Agricultural Revolution'. He was also county MP for over fifty years between 1776 and 1832; and the owner of Holkham Hall, one of the finest palladian mansions in Britain at the centre of, by far, the largest estate in Norfolk. A friend of Charles James Fox, he moved in the highest Whig social circles and lavishly entertained distinguished friends from both political and academic fields who came to Holkham for its splendid library, works of art and antiquities as well as the game coverts. A charismatic figure, he was an outspoken critic of Britain's war against the Americans in their fight for independence which made him friends who visited and corresponded across the Atlantic.

Despite his importance, both locally and nationally, there has been no full scale biography of him for a hundred years – a gap which this book sets out to address. It sets his agricultural achievements in a wider context, and places Coke himself in his milieu, as one of a small circle of landed grandees who were of major influence during a period of political turbulence and agricultural change. The author also examines Coke's reputation as a 'patriot'.

Dr SUSANNA WADE MARTINS is an Honorary Research Fellow, School of History, University of East Anglia.

COKE OF NORFOLK (1754–1842) A BIOGRAPHY

Susanna Wade Martins

THE BOYDELL PRESS

© Susanna Wade Martins 2009

All Rights Reserved. Except as permitted under current legislation
no part of this work may be photocopied, stored in a retrieval system,
published, performed in public, adapted, broadcast,
transmitted, recorded or reproduced in any form or by any means,
without the prior permission of the copyright owner

The right of Susanna Wade Martins to be identified
as the author of this work has been asserted in accordance with
sections 77 and 78 of the Copyright, Designs and Patents Act 1988

First published 2009
The Boydell Press, Woodbridge
Reprinted in paperback 2010

ISBN 978-1-84383-531-8

The Boydell Press is an imprint of Boydell & Brewer Ltd
PO Box 9, Woodbridge, Suffolk IP12 3DF, UK
and of Boydell & Brewer Inc.
668 Mt Hope Avenue, Rochester, NY 14620, USA
website: www.boydellandbrewer.com

A CIP catalogue record for this book is available
from the British Library

The publisher has no responsibility for the continued existence or accuracy of URLs for
external or third-party internet websites referred to in this book, and does not
guarantee that any content on such websites is, or will remain, accurate or appropriate

This publication is printed on acid-free paper

Text pages designed by Tina Ranft

Printed in Great Britain by
CPI Antony Rowe, Chippenham and Eastbourne

CONTENTS

LIST OF ILLUSTRATIONS

METRIC CONVERSIONS

Money
Twenty shillings is equivalent to 100 pence (£1)
One shilling is equivalent to 5 pence
Twelve pennies (d.) is equivalent to 5 pence (p.)
One guinea is equivalent to £1 5p.

Area
One acre is equivalent to .405 hectares

Volume
One quarter is equivalent to 290.5 litres
One last is equivalent to 2905 litres

Dedication

In memory of Tom Pocock (1925–2007)
who encouraged me along the way and whose enthusiasm
for Norfolk heroes was infectious.

ACKNOWLEDGEMENTS

My interest in Thomas William Coke and the Holkham estate began in the early 1970s, when it was the subject for my PhD thesis. The librarian and archivist at Holkham, the late Dr W. O. Hassall, was hugely helpful, as was my supervisor, Professor Richard Wilson of the University of East Anglia. Since then I have continued to haunt the archives there, now run by the ever-obliging Christine Hiskey, whose guidance through the material has been invaluable to me. Our many conversations over coffee and lunch will long form one of the happy memories of this research.

It was not until I embarked upon an MA course in life-writing at the University of East Anglia that I turned my attention to the possibility of writing a biography of Thomas William, and both Professors Richard Holmes and Kathryn Hughes there have provided help and encouragement. Professor Anthony Howes of the School of History helped me get to grips with the politics of the period. Other colleagues in the School of History, particularly Professor Tom Williamson, have given advice and encouragement.

The Earl of Leicester and Coke Estates Ltd have given generous financial assistance towards the publication of this book and allowed me access to the documents at Holkham, without which the research would have been impossible.

The figures were drawn for me by Phillip Judge.

I would also like to thank the anonymous reader and Caroline Palmer at Boydell and Brewer.

Most importantly, my husband Peter has as always been supremely supportive of the project.

FOREWORD

Thomas William Coke, known as 'Coke of Norfolk', is perhaps best known for his role in transforming farming, changing its almost medieval practices to those of a modern agricultural industry.

In the late eighteenth and early nineteenth centuries, the Industrial Revolution was gathering momentum; huge numbers of people flocked to the towns to find work in the new industries. It was due to Coke and his fellow agriculturalists, such as Marquess Townshend and the Duke of Bedford, that food production increased so dramatically as to be capable of feeding the rapidly expanding urban population.

Before Coke succeeded to Holkham in 1776, farming was considered a very unfashionable occupation. By example, Coke, as the owner of the largest estate in Norfolk, gave farming status. He stated, 'If you want to attract gentlemen, you must build them gentlemen's houses.' This he did, and the houses on the estate are as fine as any in the country. The farmers who came as his tenants had capital, and this, combined with long leases and an interested, sympathetic landlord, ensured modern farm buildings were erected, the latest farming machinery was affordable and up-to-date farming practices were followed.

Coke was a great self-publicist and was not unknown to exaggerate his contribution to the prosperity of agriculture. Nevertheless, his egotism was channelled in worthwhile directions, of which the annual sheep shearings are the most prominent example. These occurred over a period of three days each year from 1778 to 1821. Hundreds of people, progressive farmers, stock breeders and machinery manufacturers, descended on Holkham to view the latest advances in farming. Their observations were disseminated not only throughout the British Isles, but as far away as America. Indeed many Americans attended the sheep shearings.

However, what is not quite as well known is Coke's role as a Whig country gentleman and a parliamentarian. Dr Wade Martins draws our attention to this equally important facet of his life. He was an MP for Norfolk for over 50 years, and his interests in parliament, apart from promoting the cause of agriculture, were those of a typical Whig. He lived by the Whig motto of 'Live and let live'. He fought for Catholic emancipation and was a fierce opponent of the slave trade. His belief in the Whig value of liberty allowed him to support the colonists in the American War of Independence. He campaigned for the elimination of the rotten boroughs, the reform of parliament and the extension of the suffrage.

The passing of the Great Reform Act of 1832 signalled not only the achievement of the last of these aims but also the moment for Coke to retire from politics. Five years later, Queen Victoria succeeded to the throne. A new age, radically different from Georgian England and to which the modern reader can easily relate, was ushered in.

Not only does Dr Wade Martins's narrative guide us through Coke's long and fascinating agricultural and political life, but she also gives us an equally fascinating and intimate view of daily life through the years at both Holkham and in London. As a commentary on life in Georgian society, it is invaluable.

My very great thanks go to Dr Wade Martins for her enormous contribution in furthering our knowledge of that remarkable man, my great-great-great-grandfather, Coke of Norfolk.

The Earl of Leicester CBE, DL

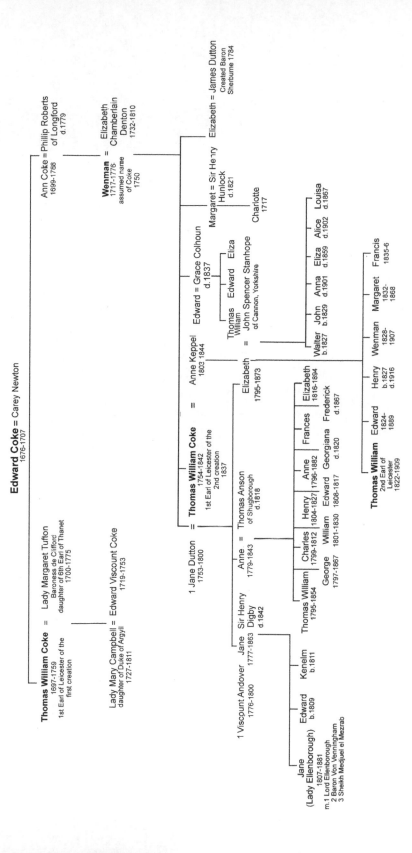

Figure 1: *Family tree of the Cokes of Norfolk. Those who were owners of Holkham are shown in bold type.*

TO BEGIN AT THE END:
THE FINAL JOURNEY

The reader must not be led to anticipate any dazzling incidents or stirring trials,
but to contemplate a shady consistent course, exhibiting the model for the life of
an English gentleman of estates and an example of a Patriot.[1]

On 7 July 1842, the hall at Longford would have awoken early. As the sun rose over the lake and the plaintive calls of the coots died away there was movement in the stable block beside the house. Four of the finest carriage horses and faithful favourites of the late earl were groomed and harnessed in their stalls under the vaulted brick ceilings. With polished bridles and leather reins decorated with black fringes and head straps surmounted by black plumes, they were led out into the yard, harness clinking and hoofs clattering on the hard cobbles. The old earl's private carriage and the coach which would carry the mourning relatives were drawn out of the adjoining carriage houses and hitched up. When all was ready they left the yard with its watching clock, built by the hall's late owner while at the height of his fame some 30 years before. Now his coffin lay, draped in crimson velvet, in the nearby church. The inscription read 'Thomas William Coke, Earl of Leicester. Died June 30th 1842, aged 88.' The coffin had rested here for two days beside the effigies of knights in armour and their decorous ladies, tombs of earlier owners of the parish. It was now transferred to the waiting hearse, drawn by six black horses, to begin its slow journey to the family mausoleum in the parish church of Tittleshall at the heart of the late earl's Norfolk estates.

This was a journey which Coke had made many times during his life and with which his horses would have been long familiar. In another July, some 71 years earlier, he had set out for the first time from Longford in Derbyshire for Holkham in Norfolk to view the estate of 30,000 acres which he was to inherit along with a house acknowledged as one of the finest examples of Palladian architecture in Britain. Much had happened since that encounter, and his reputation as a Whig gentleman of standing, an agricultural improver and, most importantly, a 'patriot' is reflected in the scale of the funeral cortège which was about to set out on this final journey.

Funerals were important occasions, in which the degree of ceremony indicated the rank of the deceased. When Coke's great-uncle, another Thomas Coke, died at

Holkham in 1759 two mourning coaches and their coachman were hired for three days, to travel the 20 miles to Tittleshall and back. Seventeen of the servants were provided with mourning suits and frocks and accompanied the chief mourners to the church.[2] When Coke's first wife, Jane Dutton, had died in Bath in 1800 at the age of 47, she too was brought back 'by slow stages'[3] to Tittleshall, where the coffin was met by tenants, but on this occasion the ceremony seems to have been a small family affair with only close relatives present. Although a simple occasion, the cost was still about £1,741, including the mourning coach (three days from Norwich), mourning suits for the servants, silk tassells for the postilions' caps and black cloth fringes for Holkham church, mansion and chapel as well as the entertainment of tenants in Tittleshall.[4] By the time of Coke's death, the etiquette of mourning had become even more elaborate, complementing the stylised ritual and impressive public ceremonies which were now expected. These added a type of glamour to ordinary life which is now associated only with great state events. A family could demonstrate its wealth and standing by the expense lavished on the funeral.

The most public expression of the dignity of the departed was in the cortège itself, the length of which varied with the consequence of the family.[5] When the Duchess of Rutland died at Belvoir Castle in 1827 her funeral procession took three hours to cover the couple of miles to Bottesford church, and when the third Duke of Northumberland died in 1847 all the tradesmen in Alnwick shut up shop and a long funeral party of tenants accompanied the coffin through the town to the station in Gateshead for its transportation to Westminster Abbey.[6] A year after Coke died his royal friend and uncle of Queen Victoria, the Duke of Sussex, was buried. His funeral included a procession from Kensington Palace to his preferred resting place in Kensall Green Cemetery through streets lined with mourners. The black-plumed hearse drawn by eight black horses was preceded by cavalry, a military band and 13 mourning coaches. Behind were the coaches of the Queen, the duke's brother, the Duke of Cambridge, and 40 private coaches of friends.[7]

The Longford tenants had gathered on horseback at the church in the early morning light. Dressed in deep mourning they led the group down the curving drive through the park and on to the road. Barely a fortnight before, they had been part of the ceremony when two new bridges financed by their landlord across the millstream which ran through the village had been opened by him amidst much celebration. The tenants were followed by the carriages of the two doctors who had attended the dead earl in his last, short illness. The undertaker rode ahead of the hearse, driven by two coachman in the front with two postilions behind, all in black, bare-headed and carrying their hats. Behind was the family mourning carriage, followed by Coke's own carriage, empty except for his valet, on whose knees rested a velvet cushion with the earl's coronet, symbol of the ennoblement which Coke claimed to have so long resisted, preferring to remain in the House of Commons and revelling instead in the title 'the greatest commoner in England'. Then came the carriages of several local dignitaries, including members of the Strutt family from nearby Belper, representatives of the new industrial class of textile mill owners who had risen to prominence during Coke's own lifetime. Since the death, in 1837, of his brother, Edward Coke, who had lived at Longford, Thomas had spent more time there and had built farmhouses, almshouses and a school. The procession left Longford along the country roads, first winding steeply and then a few miles from Derby, following the straight Roman road flanked by the enclosed fields of the improved agriculture for which the Cokes were famous, were lined by midday with the crowds who had gathered to see the spectacle. The fast-growing county town with its stocking knitting and porcelain manufactories was strangely silent. The large market-place was deserted and the shops closed. The only sound was the tolling of church bells. Shortly after Derby the Longford tenantry and local carriages turned their horses westwards and headed home into the late afternoon sun, having bid a final farewell to their landlord and fellow landowner of 66 years.[8] Very few of them would remember Coke's father, Wenman, yet the evidence of his dedication to the welfare of his estate was obvious for all to see. He it was who had first enclosed the fields and encouraged the new methods of farming with which his son's name was to be so closely associated. His Home Farm next to the new carriage house had been run on an exemplary pattern. Thomas had continued to take an interest in the farm. The buildings, erected in 1760 and 'very convenient for the inwintering of cattle',[9] were an easy walking distance from the hall, even for an old man.

From Derby, the cortège travelled on to Nottingham, an expanding industrial town famous for lace and stocking knitting, but one with which Coke had had far fewer connections so the procession did not stop there. Instead it moved on to Bingham, reached by eight o'clock, where the coffin lay in state in the Chesterfield Arms, an inn standing between the church and the market square. On the second day the procession followed the familiar route through Grantham and on across the limestone uplands, staying the second night at the Wheatsheaf, opposite the large, ornate church in Swineshead on the edge of the fens, famous as the site of the monastery en route to which King John lost his treasure. Entering the fens across the Roman Foss Dyke, the carriages travelled through a different world. The mourners looked down upon the flat marshland peppered with the turning sails of drainage mills and the occasional wisp of smoke from a steam pumping-engine chimney: a scene of agricultural improvement, much of which had taken place since Coke had first made the journey. On the third day they crossed the embankment and bridge at Cross Keys, erected in 1830, into Norfolk, Coke's adopted county, which he had represented for most of his long parliamentary career stretching from 1776 to 1832, with a break of only six years.

From the point of entry into Norfolk, the crowds had begun to gather 'amid deep silence, only interrupted by the tramp of the horsemen, the grating of the carriage wheels, and the heavy tolling of muffled bells in the villages through which it passed'.[10] Crossing the bridge and passing the tollgate, the slow procession skirted the marshes along the Wash, through the prosperous, straggling fen villages with fine church towers, their height accentuated by the flatness of the land around them, and standing as symbol of the early and continuing wealth of this watery region. Against the sky, their naves appeared transparent as the sun shone through the clear glass of the large traceried windows. In Long Sutton the inhabitants 'testified their respect for the deceased earl by partially closing their shops and the minute bell tolled'.[11] To the north, between the road and the sea, sheep and cattle grazed on the newly drained estates of Lord George Bentinck, now Kings Lynn's MP and a supporter of Coke's sworn enemies, the Tories. A patron of Disraeli, he belonged to a different generation from Coke, who had been most active in the days of Pitt and Fox. The cortège turned south to the new bridge across the Ouse, where only a ferry had existed in Coke's early days, and so into Kings Lynn, through the medieval Southgate with its turrets draped in black flags. The whole town was in mourning as the procession entered. Black flags fluttered from all the public buildings, including the fine old guild-hall with its knapped flint façade. All the boats in the harbour were also flying mourning pennants. The shops were closed and many of those waiting in the streets to watch the coffin pass were dressed in black. That night the coffin lay at the Crown Hotel[12] in Church Street, where 1,500 people filed past, and 'owing to the lateness of the hour more than a further thousand respectable individuals were turned away'.[13]

On the next day (Sunday) the coffin was moved on through the prosperous countryside to the river port at Narborough, the last point to which wherries could bring coal inland from Lynn. From there it travelled across the corner of Swaffham Heath, to the thriving market town of Swaffham, with its elegant, domed butter market, given by another Norfolk Whig and dignitary, Sir Robert Walpole. On the morning of Monday 11 July, five days after the coffin had left Longford, the final and most extraordinary phase of the journey began in bright sunshine ('As if to bear testimony to his virtues, the sun was permitted to shine on this, his last day of earthly honour.').[14] A long funeral procession formed, led on horseback by William Baker, the Holkham estate steward, followed by the tenantry 'headed by the senior tenants, most of whom are known to be amongst the most skilful agriculturalists in the world'.[15] Behind them came the undertaker on his horse, followed by the hearse. Then came the carriages, carrying the clergy (one of whom, the Revd Kenelm Digby, was Coke's grandson and rector of Tittleshall), the pall bearers, all members of the local gentry, the executors of the will and the upper servants. Coke's open coach, brought over from Holkham specially, with his valet and the coronet followed. Then came the hearse itself with its six horses, followed by the many mourning carriages, each pulled by four horses. Finally in this part of the procession came Coke's own carriage with its shutters closed. Behind this formal cortège came the carriages of friends, then 200 men on horseback riding two abreast and finally a long train of neighbours. The procession was already two miles long as it started and it is hardly surprising that such a spectacle should attract large crowds. It was said that 150,000 were assembled along the route near Swaffham, many of whom had camped overnight in fields from which the hay had just been harvested, to get a good vantage point. At every crossroad, more mourners joined the procession, which, by the time it reached Tittleshall, was said to stretch back to the crossroad village of Litcham, some three and half miles distant. The road was a 'moving mass of beings' which made it difficult for the procession to pass.[16] If the length of the cortège was an indicator of standing, then Thomas William Coke of Holkham, first Earl of Leicester, must have been rated as highly as any royal duke.

On arrival at the Ostrich public house in Tittleshall (the ostrich featured on the Coke family crest and gave its name to pubs throughout the estates) the mourners dismounted and walked the short distance to the church. It is difficult to imagine the scene in this small mid-Norfolk village on the edge of which the founder of the family's fortune, the great sixteenth-century Lord Chief Justice had built his hall and where the Cokes had since been buried in the brick mausoleum on the north side of the church. Over a hundred carriages and 400 horses were parked and tethered in the fields around the pub, the rectory and nearby yards of the farming tenants. The church would have been more than filled for the service conducted by Mr Digby and his curate, the Revd Mr Dashwood. In the chancel was the splendid monument to Coke's uncle, an earlier Thomas, by Messrs Atkinson of London, the contractors for all the marble work in the hall, and commissioned by a grieving widow nearly 90 years earlier. On it she placed two busts, one of Thomas and one of herself by the fashionable sculptor, Roubiliac. Nearby, another monument commissioned by Coke from the famous Joseph Nollekens, in memory of his first wife, Jane Dutton, depicted a Classical scene with a sorrowing woman, believed to be a portrait of their daughter, also Jane (see page 70).

The coffin was then carried into the mausoleum 'beside the tears and sighs of a deeply affected assembly'[17] and for several hours the public were admitted through the small vault door to view it, as it lay covered with red velvet decorated with two borders of brass studs, the coronet lying on it. Finally it was inserted in its niche beside that of his first wife, and the spectacle was over (Colour Plate 1).

What happened next is not recorded. After the funeral of his great uncle, the first Thomas, 10 guineas had been distributed among the poor of Tittlehall, but there is no record of such largesse this time. The cost of entertaining the tenantry would have been covered by the young new earl and his family. The Ostrich would certainly not have been able to cope with such crowds. The rectory opposite the church, though large, could have accommodated no more than the immediate family, and the local farmers could have provided hospitality for their fellow tenants in their fine large houses, mostly faced in distinctive Holkham white brick and remodelled or rebuilt over the previous 50 years at Coke's expense. No doubt it was late at night before the village returned to its normal quietness and the last carriages and horses clattered away.

What was it that brought the crowds out in such unprecedented numbers? The weather probably helped, as did the fact that early July was a quiet period on the farms, between the hay and grain harvest. Even at a time when such spectacles were rare, this immense degree of interest is remarkable and can only be compared with that shown in the funerals of the grandest of the nobility.

Some of the explanation can be sought in the fact that Coke of Norfolk lived so long and through such a period of change. 'It was this very circumstance; this happy length of years, this unusually protracted period of usefulness, which so endeared him to us.'[18] It is unlikely that any tenant who attended that funeral could remember a previous landlord. His passing marked the end of an era. During his lifetime Britain had changed from a predominantly agricultural to an industrial nation, and some of that change can be illustrated in the journey between his Longford and Holkham estates that Coke so often took over the years. In 1771, when he had first visited his great-aunt, the route would have been far more circuitous. East–west routes were always difficult; there was no bridge at Lynn or across the Ouse at Sutton Bridge so the coach traveller would have had to go south as far as Ely before turning east for Downham Market and on to Lynn on the east side of the river. New turnpikes were being opened all the time, and gradually the journey would have become easier. The towns of Nottingham and Derby through which he passed changed beyond recognition during these years, with mills powered both by water and by steam springing up beside rows of newly built houses. Nottingham's growth in particular was restricted by the lack of the enclosure of the surrounding open fields so the houses became more densely packed back to back as an increasing population was housed around yards. Whilst the town had enjoyed a reputation as an attractive centre for county gentry in the eighteenth century, by 1845 its slums and overcrowding were said to be worse than in any other industrial town. The growth of industry and the wealth it created for some is demonstrated by the fact that the Strutts, who had built their first mill in Belper on the Derwent valley north of Derby in 1787, had, by 1842 increased their social standing to such an extent that they accompanied Coke's coffin through the county town.

Agriculture, too, had changed dramatically and it was with this that Coke's name has always been linked. His obituary in the *Norfolk Chronicle* stated that it was 'as an agriculturalist second to none that he will be most remembered'. In the words of his memorial at Longford, 'he extended in a remarkable degree the cultivation and rural improvement of the country'. The publicity he gained through his annual sheep shearings goes some way towards explaining this reputation. On his estates it was his hospitality and liberality, his lack of affectation and impatience with ostentation which gave him the reputation of a warm and sincere man.

But Coke saw huge political as well as industrial and agricultural changes of which he too was a part. His role in public life was always mentioned alongside his agricultural exploits. He was described as a 'great patriot and benefactor of mankind',[19] whose aim had been 'the amelioration of the civil and political condition of his fellow men'.[20] He entered politics at a time when the Tories were still a discredited group associated with the Jacobite cause and political divisions were grouped around families and factions within the Whig party. The issue of American independence dominated much of the first few years of his political career, followed by the French Revolution, when Coke associated himself with Fox and support for all the Revolution stood for. By the time he left parliament some 60 years later, alignments had changed. The Tories, firstly under Wellington and then under Peel, were a force to be reckoned with. Whigs were becoming Liberals and the prime minister, Lord Grey, acknowledged that his 'supreme achievement' was to see the first Reform Act passed, which signalled the moment for Coke to retire.

It is necessary to look beyond Coke's work for agriculture, in which he was following in the tradition of his father and great-uncle and which was typical of many other landowners of the time, for some explanation for his reputation. In politics, whilst he mixed with the great and was a friend of Fox, there were few occasions on which he made significant speeches or contributed to debate and he never held a position in government, even when Fox was for a short time in power. Rather, his fame was based on his larger-than-life personality that led to his nickname of 'King Tom' amongst his parliamentary colleagues.[21] It was his 'kindness of heart, affability, unostentatiousness, liberality, hospitality and constant cheerfulness' alongside his 'strict integrity over seventy years of public life' that meant he was 'idolised in his neighbourhood'.[22] All these are the qualities associated with the 'old English gentleman' and 'patriot', images which he cultivated to the full, and help to explain why the crowds turned out in such numbers to see his coffin pass, making it 'one of the most numerously attended funerals ever known in Norfolk'.[23]

The term 'patriot' was frequently used to describe Coke and his brand of politics. It was a word used in the Hanoverian eighteenth century when Britain was almost constantly at war with France and describes that Whig confidence which saw Britain under the newly united crown resting on the twin pillars of Protestantism and liberty. Britain was a land of freedom and commercial prosperity, 'secure and confident under the leadership of wise ministers and kings, sure of its superior place in Empire and the world'.[24] Patriotism was seen as very different from 'nationalism', a term not invented until the mid-nineteenth century. Yet this Whig connotation was not without its critics. While in the 1755 edition of Johnson's *Dictionary* 'patriot' was defined as 'one whose ruling passion is love of his country', 20 years later he had revised the entry to 'a factious disturber of the government' and 'the last refuge of the scoundrel'.[25] This change was the result of a radical shift within a section of the Whig party led by Charles James Fox. To them the word had a far wider meaning than simply a love of one's country: to them the patriot could oppose the 'corruptness of government' and the 'tyranny' of George III. The Revd Richard Price, in his sermon published in 1789, described the 'true patriot' as a citizen of the world with a duty to 'enlighten' and liberalise rather than one with a vulgar conviction of the superiority of his country.[26] Under this definition Fox and his followers could support the colonists in their fight for independence and the French Revolution with its idea of liberty, and still be classed as patriots. However, as the French Revolution turned from a liberal idealist's dream to a war for national survival, so the meaning of the term narrowed, with more pro-establishment associations in which not only attachment to the monarchy and Empire, but the value of military and naval achievements were all-important.[27] A strong and stable government dominated by a landowning elite was seen as crucial to military success. Coke, too, was able to shift his position to suit the wartime situation.

Vicesimus Knox, headmaster of Tonbridge School, defined a patriot in his essay 'The Idea of a Patriot' (1784) in a way that comes closest to describing what Coke's contemporaries meant by the term and what Hanoverian values they saw dying with him: 'Every good man is indeed a patriot', wrote Knox, 'for a good man is a public good … the truest patriotism in not to be found in public life (but) he who secretly serves his country in the retired and unobserved walks of private life. His motives must be pure and he is a patriot.'[28] Patriotism in its late Georgian sense did not imply insularity, and the education of a patriot often included travel in Europe and participation in the Grand Tour. A patriot should be closely involved in the affairs of his country both at local and national level, and it is as an example of this particular type of wealthy liberal and hospitable Whig rooted in the gentry, rather than as a man holding a unique and extraordinary position in the history of agriculture and politics, that Coke is considered in this biography.

THE MAKING OF A PATRIOT: COKE'S EARLY YEARS

Thomas William Coke was born in London in 1754 to Wenman and Elizabeth Coke of Longford Hall, during his parents' annual migration from their home in Derbyshire to the capital for the parliamentary session. Wenman represented Derby as Whig MP for many years, and rented a house in Hanover Square, which in the 1750s was on the edge of London, surrounded by fields. One of Thomas's earliest memories was being held up to a window to watch a fox being cornered and killed by hounds belonging to an Essex pack.[1] Wenman had inherited the Longford estate from his uncle in 1750, and was responsible for making many changes and improvements there. These were two of the strands that went to make up Thomas's life, both typical of a small elite of wealthy country gentlemen of his time: the London court and politics, coupled with country pursuits often centred on agriculture and the improvement of estates. At Longford he would have first learned the pleasures of country life and to take an interest in agriculture and estate management, while in London he would have been introduced to the society and conversation of the Whig magnates entertained in the Hanover Square house.

Thomas's mother was Wenman's second wife, his first having died six months after the death of their only child. Elizabeth was the daughter of George Chamberlain of Hillesden, Buckinghamshire and through her he inherited the Hillesden estate, finally sold in 1823. Thomas was the eldest son and his birth was followed by that of a brother, Edward, and two sisters, Margaret and Elizabeth, of whom we hear very little. In 1769 Margaret married Sir Henry Hunloke and little more is known of her except that she was a frequent visitor to Holkham and died in 1821 'after a period of mental derangement'.[2] Coke's younger sister, Elizabeth, married his friend, James Dutton of Sherborne in Gloucestershire. Later, Coke married James' sister, Jane, bringing the two families even closer.

Wenman is rather a nebulous figure. The only description we have portrays him as a retiring person who 'saw little company and lived much out of the world; his habits were those of a country gentleman, bending his mind to agriculture, moderately addicted to field sports and more than either, to reading in which he passed many hours; firm in his principles which were those of the old Whig: amiable in his disposition mild in his manners, he was beloved of his friends'.[3] It is clear where Coke's interest in agriculture and field sports came from, but his own ebullient personality was in great contrast to that of his more studious father. When Wenman inherited the Holkham estate in 1776 the *Norwich Mercury* described him as 'independent-minded, benevolent and a stout advocate of the Constitution'.[4] The only public reference that Thomas Coke made to him was in his farewell speech to his constituents, when he recalled two pieces of Wenman's fatherly advice: 'never trust a Tory' and 'stick to friends and disregard enemies'. His work on improving his Derbyshire estates was admired by the Suffolk gentleman farmer and agricultural commentator who later became secretary of the Board of Agriculture, Arthur Young. He described Wenman as one of the 'most spirited farmers in Derbyshire', intent on introducing the 'Norfolk husbandry' to the county.[5] He appears therefore to have been a conscientious landlord and adequate Member of Parliament.

When he was five years old, the young Thomas's future prospects altered radically with the death without a direct heir of his father's uncle and builder of Holkham Hall, another Thomas Coke and Earl of Leicester. Born in 1697, Thomas was left an orphan at the age of 10 and brought up by guardians. He enjoyed one of the most lengthy and well documented of eighteenth-century grand tours of Europe, lasting six years. It had a huge influence on the house he built at Holkham. He returned with Classical treasures, fine paintings and a large library. His main interest, however, was in architecture and he received lessons from some of the leading Italian architects of the time. He met William Kent and Lord Burlington and together they began to plan the Palladian masterpiece which Thomas would build on his Norfolk estate.

Back in England, he entered politics as a staunch Whig and supporter of Sir Robert Walpole, receiving sinecures and public appointments under his government. Unlike Sir Robert, who sold in time and so made a huge profit, Thomas lost heavily in the South Sea Bubble, which delayed his building schemes at Holkham. In 1733 he received his most lucrative appointment as joint Post Master General with a salary of £1,000 a year. In 1728 'in consideration of his great merits' he was advanced by George II to the peerage as Lord Lovell of Minster Lovell in Oxfordshire, an estate which he owned, but was later sold by his successor. Sixteen years later he was raised still further and made Viscount Coke of Holkham and Earl of Leicester.[6]

In addition to his involvement in politics, he was also busy improving his estates, and here again young Thomas would follow in both his father's and great-uncle's footsteps. He was a man with a temper, described by Horace Walpole as a 'very cunning man but not a deep one', who owed his preferments to Sir Robert Walpole, who 'patronised him for his great estate and for being so near a neighbour in Norfolk'.[7] The descriptions of craftiness and boisterous masculinity which are found in the writings of Horace Walpole and others do not fit easily with this thoughtful and erudite man, responsible for the building of Holkham.

The circumstances of his death are not clear. It was recorded as 'sudden', and occurred six weeks after he had been challenged to a duel by a neighbour, George Townshend, over the killing of some of Townshend's foxes by Lord Leicester's gamekeeper. This was followed by insulting remarks about the local militia, of which Townshend was first colonel. Leicester refused to fight, but rumours that a duel had taken place abounded and a pistol shot may have been the cause of his death.[8] However he died, his demise ensured that on the death of his wife, Margaret, Wenman would inherit the large and valuable Norfolk estates.

The dowager Countess of Leicester had every reason to be a bitter woman, resentful of the fact that she had no son to inherit the house which her husband had so painstakingly created. Of several children born, only Edward, born in 1719, survived infancy (Colour Plate 2). He had been a promising young man, attending Westminster School and Oxford. A firework display and ball celebrated his coming of age, and he was elected MP for Norwich. However, a lifestyle that would eventually kill him was obvious from an early age. His father frequently had to pay his gambling debts and he was drinking heavily. A marriage which was to prove disastrous was arranged with Mary Campbell, the daughter of the Duke of Argyll. She was rejected by her husband on her wedding night: he preferred his drinking and gambling companions to his wife. Both parties treated each other with equal disdain and their behaviour was notorious in fashionable circles. While Edward was dissolute and disreputable, Mary was volatile, with a passion for the sensational, and her portraits show her with a small, imperious face. Mary was by all accounts held a virtual prisoner at Holkham for a year before suing for divorce. The publicity given to the case was enormous, the proceedings attracting huge audiences. However, Lady Mary's claims of cruelty by her husband were far too vague to stand up in court and so the case failed. Finally she agreed to live quietly with her mother in Sudbury, which she did until the death of Edward in 1753, when she again entered London society. In her full and lively journals she recounts both her travels in Europe and life in London, where she rented a house neighbouring the fashionable Holland House, home of Charles James Fox's nephew and friend of Thomas William, Lord Holland. She carried on a long and intensely emotional flirtation with Edward Augustus, Duke of York and brother of the future George III. Twelve years her junior, he probably did not return her affections, and the affair was regarded as faintly ludicrous by contemporaries. She was also a close friend of Horace Walpole, who, while not blind to her many faults, frequently defended her against critics and extricated her from her misadventures. With her widow's annuity of £2000, she could afford not to marry again. She finally died in 1811.[9]

With no direct heir to follow her, Lady Leicester set about finishing the work on the hall which was uncompleted at her husband's death, while vowing to outlive her husband's nephew, thus denying him the chance of inheriting what should by rights have been her son's. There is no evidence that Wenman was ever invited to

Holkham. Even after it became clear to him that he was the heir to Holkham he studiously ignored that branch of the family, to the extent that the dowager at Holkham accused him of great discourtesy. Writing to her friend Charles Yorke in 1767, she reported, 'Our County Meeting is over; Mr. (Wenman) Coke being nominated a [parliamentary] Candidate surprised me, as he sent me no notice of his design' – an omission that would have seemed very rude to the titular head of the family[10] and led her to believe that he had come to 'nose her in her own county'.[11] Her dislike went with her to the grave, and in her will she declared austerely that Wenman's conduct towards her had not been such as she had just reason to expect, and that if he had treated her better, he would have had some of the benefits which she now bequeathed to her sister.[12] There is no record that they ever met. In contrast, she treated her steward, Ralph Cauldwell, very favourably, which led to years of wrangling when young Thomas finally inherited.

Very little is known of Thomas William's early years. The Revd Thomas Keppel's manuscript biography at Holkham, on which much of what follows relies, states that he started his schooling in Longford village before going on to a fashionable establishment run by a French refugee in Wandsworth.[13] In 1765 he was sent to Eton, just as his friend of later years, Charles James Fox, was leaving. Soon he would be joined by future neighbour and lifelong friend, William Windham of Felbrigg, who was four years old. Eton at this time was the centre of a very small political world. In 1764, its headmaster was Edward Barnard, who, though not perhaps as scholarly as other headmasters of the century, was an able administrator, which was more important. He was popular amongst his students, to whom he taught Greek plays which the sixth form performed. Under his management the school flourished and grew from 300 to 552 pupils, many of them aristocrats.[14] However, in 1766 he was followed by John Foster, a small man with poor eyesight and the son of a Windsor tradesman. All these factors worked against him. He also lacked any sense of how to manage boys, and was so unpopular with the pupils that there was a rebellion in the school during which about 170 of the senior boys walked out, books were thrown in the river, some masters' windows were broken and a couple of masters manhandled.[15] Many boys left, and some, including William Windham, were sent home. A letter from an undermaster and legal guardian of the young Windham, Dr Dampier, to his mother explained that her son was accused of having a 'large concern' in the riots and so 'In order therefore to cover his retreat and to prevent a public expulsion, which would probably be a consequence of a longer stay, I shall see him home to you tomorrow morning.' Windham was renowned amongst his fellow pupils as a fighter, cricketer and oarsman, and so was likely to be a respected leader.[16] We do not know whether young Thomas was involved in this mass insubordination, but from what we know of his independent spirit, it is quite possible, although, unlike Windham, he did not leave school until 1771. By 1773, the numbers in the school had halved and Foster resigned.

As well as the headmaster, there were about a dozen teaching staff and those who ran the boarding houses. Boys were grouped for work by ability rather than age, which often led to bullying by older boys of their younger, if brighter, fellow pupils. Fighting was the usual way in which disputes were settled. There is no indication that Coke was unhappy under this regime, and indeed he seems to have worked the system to his advantage. Study was limited to three terms, with a month's holiday at Christmas, two weeks at Easter and a month in the summer. Academic work consisted of the study of the Greek and Roman authors, with older boys required to compose two Latin exercises, one in verse and one in prose, every week. They were expected to read history and literature in their leisure time, but from what anecdotal evidence survives about Coke's school-days, he can have had little time for such activities.

Instead his main interests were in field sports, which meant he was popular: 'his gun would provide supper for his school mates.'[17] He could persuade the more studious boys to undertake his Latin homework in return for game. The story goes that on one occasion he was found with 70 snipe in his room that he had shot, and that he narrowly escaped more serious punishment for killing a pheasant in Windsor Park. Flogging was the normal method of discipline, and Coke and his companion were flogged by the headmaster for this misdemeanour. It is quite clear that Coke took very little interest in academic pursuits during his school years and the books left to the school by Sir Robert Topham in 1730 to be housed in the new library would have been little used by him, although amongst them would be titles to be found in the fine library collected by his great-uncle which would be his when he inherited Holkham.

During the school holidays at Longford, he would be out all day with his gun, and by the time he left school at eighteen he was a tall, strong, active young man. He was also a very good sportsman with a smattering of the classics and friends and connections in the small world of the landowning elite which would stand him in good stead in future life. His knowledge was that most prized by country gentlemen involved with the management and improvement of their estates:'practical rather than book knowledge'.[18] These are attributes that lasted all his life and shaped his actions through the following 60 years.

There is no reason to doubt that he was present at the wedding in 1769 of his sister Margaret to Sir Henry Hunloke of nearby Wingerworth, Derbyshire. After the 50 servants were settled in the church, the family processed there from the great house, Sir Henry dressed in a white satin sequined suit. Back in the house the ceremony was performed again 'in the Romish manner', and it may well be that this family link with the Catholic Church was one of the reasons why Coke was such an enthusiastic parliamentary supporter of Catholic emancipation. There followed toasts and a wedding cake, some 'cut into thin slips and drawn according to due form, through the wedding rings for the use of many unmarried friends of both sexes, put in clean paper and sealed and directed properly'. Everyone then dressed for dinner at three, which included fantastic pastry and sugar creations by the cook. The following day the bride and groom dressed in their wedding clothes again and went for a repeat ceremony in the Roman Catholic church in Derby.[19]

Coke's first journey after leaving school in the summer of 1771, which was to have great significance for his future, was to Holkham. The options for the completion of the education of young men on leaving Eton were simple. They either went to one of the ancient universities or they travelled on what was known as the Grand Tour. Normally accompanied by a tutor, their route would take them through France to Italy, as well as sometimes to Switzerland and Germany. By 1700 it was established as the ideal way in which to impart both taste and knowledge to a young gentleman and was undertaken by many of Coke's future Norfolk neighbours. Normally the tour lasted about two years, allowing time for the traveller to become a connoisseur of art and gather a collection reflecting his own interests and taste. In this way the much-admired works of art at such Norfolk houses as Narford, Melton Constable, Langley, Felbrigg, Houghton and Wolterton were assembled. Coke's great-uncle had been unusual in spending six years abroad, between 1712 and 1718, studying architecture and collecting the books, pictures and statues which were to adorn his new house.[20]

Given Coke's lack of interest in things academic, it is unlikely that his natural choice would have been to go to university. However, the decision was almost made for him. His great-aunt, the dowager Lady Leicester, offered him £500 towards his travels if he did not go to university. These she regarded as 'schools of vice' and she saw her own son's university days as leading to his disgrace and premature death. Wenman Coke agreed to add another £200 to his aunt's £500 and Thomas was ready for his travels. This would probably have been enough to cover his expenses, although the cost of the Grand Tour could vary wildly. While young Coke's great-uncle Thomas had been allowed £1,000 a year by his guardians,[21] £300 a year was considered 'full sufficient for any private gentleman' in 1760. In contrast, the second Earl of Fife managed to spend £1,700 in a few weeks in Paris in the winter of 1766/7.[22] Before Thomas went, however, it was necessary to visit Holkham and the old lady there.

So it was that in July, almost exactly 70 years before his coffin made its final journey, Thomas William Coke set out along the Derbyshire lanes to visit his great-aunt. A youth spent out shooting across the fields, listening to his father and possibly visiting his tenants and attending the annual audit dinner must have created an interest in what he saw through the carriage window. As well as the neat enclosed fields of his father's estates, there were the open fields of much of Nottinghamshire, where tenants holding a variety of strips scattered across several open fields still farmed in traditional ways. Here in July he would have seen fields that had been left fallow all year, whilst on the improved farms of Longford there would have been the lush green leaves of rows of turnips, providing a break crop between cereals as well as fodder for livestock in winter, thus removing the need for fallow, regarded by improvers as 'wasteful'. As well as the open fields were the heaths where stock were kept during the summer. In a dry year some of these commons would be looking rather brown by now as many were over-grazed.

This route by coach or on horseback along poor roads would be improved during Coke's lifetime by the turnpike trusts. The road from Kings Lynn to Fakenham and on to Wells was turnpiked between 1826 and 1828, while the route to London via Ely was already improved some 12 years before Coke inherited. The railways, which in so many ways signify the change from Hanoverian to Victorian Britain, did not reach Norwich until five years after his death. Holkham itself was not connected by rail to the market town of Fakenham until 1866, with a line largely financed by Coke's successor.

As the traveller entered Norfolk across the fens, he would have seen that much land was still too wet to graze, except in the summer, when sheep and cattle were turned out on rushy and weed-infested meadows. Leaving Lynn for Fakenham on the final phase of his journey, he crossed common and heathland, some of it so sandy that it was mostly left to rabbit warrens. Fields were both open and enclosed. Beyond the enclosed fields of Gayton were the open fields of Gayton Thorpe and the common of Massingham. Here Thomas caught his first sight of the land which he was to inherit. The ownership of Massingham was divided between the Walpoles and the Cokes.

As Coke was travelling to Holkham, the agricultural writer Arthur Young was also visiting the region. He called on Mr Carr, the Holkham tenant of a Massingham farm. Young was impressed by Carr's progressive farming, and noted the newly enclosed fields which had be treated with marl – the chalky subsoil found across this area of Norfolk – which could be dug up and spread on the lighter acid topsoils to improve their quality. Mr Carr was not only an expert in the use of marl, but also on the feeding of oil cake (the rich residue left when oil had been extracted from oilseed rape) to cattle. In July both the cattle and the sheep would have been out in the fields. The sheep on Mr Carr's land would have been folded within hurdles on the grass, whilst come winter, they would be on the turnips. Four hundred of Mr Carr's sheep would eat an acre of turnips a day. Only a few years previously the whole area was sheep walk, with very little in the way of crops.[23] The Carr family was one of many on the estate who would be encouraged by the future owner now travelling through his inheritance, and who would help enhance the reputation of Holkham as the cradle of improved agriculture.

Turning north after the prosperous village of Litcham, he would soon have entered other parishes which were outliers of his future inheritance (see map page 83). Tittleshall and Godwick were at the heart of the seventeenth-century family possessions, and to the east were Wellingham and Weasenham, bought by the Chief

Justice in the 1600s. Weasenham had also been visited by Young in his search for progressive farmers. He had interviewed Mr Billing, well known in farming circles for the quality of his carrots and turnips[24] and a member of another family that was to receive encouragement from the future landlord. Further north were the lands of the Townshends of Raynham. Charles Townshend had gained the nickname 'Turnip' Townshend for his interest in improving agricultural practice. He had died in 1738, but the appellation has been with him ever since. Both the enclosed fields and the parkland of Raynham Hall with its clumps and shelter belts so suited to the protection of game birds would have met with Coke's approval. Past Hempton Green was the small coaching and market town of Fakenham, where there were several inns to choose from for a much-needed rest.

The last stage of his journey from Fakenham to Holkham took him into the chalky soils of north-west Norfolk with which he was to become so familiar. From Fakenham to Walsingham the road passed through enclosed country, much of it owned by the Astley family, with whom he was to be closely associated in his parliamentary career. Beyond Walsingham, now a sleepy village of timber-framed houses, but before the Reformation second only to Canterbury as the most popular place of pilgrimage in England, he entered the main area of the Holkham estate. Mostly enclosed for cereals, but with areas of heath remaining, it must have been obvious to a young man brought up with a farmer's eye that this was good land let in large productive farms which, given the right calibre of tenants, would ensure that Holkham was a profitable estate.

Finally, after four or five days of travel, Keppel describes in his manuscript biography, probably dictated by Coke in his old age, how the weary and rather nervous teenager[25] entered the park of the dowager countess under the triumphal arch leading to the south gates. From here, his eye would have been taken straight up the hill where, at the highest spot in Norfolk for many miles around, stood a stone obelisk, its needle-like point reaching skywards. The trees on either side of the avenue leading to it were neither thick nor tall enough to hide the farmland and fields lining the way. He would have noted with some disappointment the lack of clumps, so important for game cover. The obelisk itself was surrounded by a plantation of young trees cut through to create radiating rides with views of distant buildings such as a small Classical temple and, at a greater distance, the church. It was not until the carriage rounded the obelisk that the south front of the great Palladian mansion came into view. Built of yellow brick, to emulate Palladio's villas around Venice, and with a slate roof, its great core was flanked by four identical wings which terminated in pavilions. In front of the house was a formal canal, part of the landscape designed when the house was begun in the 1730s. He would have caught a glimpse of the lake as the coach turned round the side of the house to the main entrance on the north. Beyond were the open salt marshes and the sea, where boats waiting for the tide to go up to Wells were anchored. The lake was nearly straight and the dam did not always hold the water in, so it could be reduced to a muddy creek: something to be improved in due course (Colour Plate 3).

Whatever he had heard about the house, nothing could have prepared him for the sight that greeted him as he climbed down from the carriage to be met by the porter and servants in their state liveries. Lady Leicester had decided to make every effort to impress.[26] It was not the row of servants that took his breath away but the great entrance hall in which they stood, surely the most impressive of any house in England. In sharp contrast to the simple, small door through which he entered was the grandeur of the hall. If he had used the library at Eton he might have recognised the similarity between the room he had entered and the basilica-like banqueting hall of the gods described in Ovid's *Metamorphoses*. The lower level was entirely lined with pink Derbyshire alabaster, a sight that 'quite shocked' the French brothers La Rochefoucauld when they visited some 10 years later.[27] The marble extended to a gallery level, off which the state rooms opened, flanked by rows of pillars with gilded Ionic capitals copied from the temple of Fortuna Virilis. Rising into the apse at the far end was the huge staircase, leading the eye up to the large mahogany doors opening into the grand saloon, and then on, to the sumptuous plasterwork ceiling modelled on that in the Pantheon in Rome.[28] Bands of flowers and leaves framed trapezia of differing sizes to enhance the perspective within the apse. Above the saloon door was a ledge intended for a head of Jupiter which Thomas Coke had brought back from Rome, but instead his widow had placed a fine Classical bust of her late husband there, thus honouring the creator of this architectural masterpiece. Not completed until seven years before young Thomas's visit, with the gilding still glistening, the whole design bore the hallmark of the Grand Tour and the Classical civilisation his great-uncle had studied in such detail and admired so much. From the moment he entered the house, the visitor knew that this was the creation of a man of culture and impeccable Classical taste.

It was not to the saloon that Coke was led to meet his great-aunt, but around the gallery looking down into the body of the hall, along a corridor with windows opening into one of the central courts and through a state bedroom to the square, south-facing sitting-room which served as an ante-room to the dining-room and, beyond, the saloon itself. Nervous and travel-worn he may have been, but even at 18 he was a tall, impressive figure, well-built, with a strong nose and dark-brown hair. His first meeting with the dowager is described in detail in a second manuscript biography at Holkham, possibly written by Coke's admirer and editor of the Whig *Norwich Mercury*, Richard Noverre Bacon. This too may have been dictated by Coke himself, and describes the small, delicate lady who came in to greet him and sat beside him on the sofa. She made it plain as she clenched her fist that although he was to inherit the estate, she intended to live 'as long as I can'. With these words she left. Whether it was to the strangers' (or guest) wing or to rooms in the family wing that young Coke was escorted after his audience to change out of his travelling clothes is not clear, but later, at nine o'clock, he sat down in splendid isolation to a formal supper at which he was served by a full retinue of servants.[29]

His visit to Holkham lasted about a month, during which he was treated with 'icy civility'. Meal times had to be kept to the minute, and there was little contact with anyone but his great-aunt. His time was spent exploring both the interior of the great house and the park around it. Occasionally he got lost within the house, and he remembered his great-aunt's anger when he was late for breakfast as a result. The house, which covered an acre of ground, consisted of a central block containing the state rooms and four identical pavilion-style buildings at each corner. The north-east one was the strangers', or guest, wing. Although more plainly furnished and decorated than the family wing to the south, it was still the size of many a small gentleman's-residence, with well-proportioned rooms. From here a corridor led into the north 'tribune' or ante-room of the great statue gallery which occupied the whole eastern side of the main block. With its west-facing windows and light walls to show off the Roman statues to their best, the room was flooded with evening light. If the opposing east and west doors of the ante-room were open, then it was possible to see through the north dining-room, across the great hall to the state sitting-room and, beyond, the north state bedroom, which was occupied by the dowager. She had chosen this room because she could keep an eye on the comings and goings of her guests. His lonely wanderings around the house would not only have included the huge, richly decorated and formal state rooms, but also the less daunting rooms of the family wing, with bedrooms, a south-facing sitting-room looking out over the canal to the obelisk which he was to make very much his own and, at first-floor level, the library, in which the books collected by his great-uncle were shelved in the fine gilded cases designed for the room by William Kent. The chapel wing contained both the light and airy chapel and empty rooms awaiting the children, their nannies and tutors for which it had been designed but which were never used. It was now mainly occupied by the upper servants. The kitchen wing to the north-west contained the vast kitchen and further servants' quarters. In this wing were kept the boxes of manuscripts which were to excite his academic friends in the years to come.

Though he might have had the use of a horse for excursions a little further afield, contact with those he passed on the way was firmly discouraged by his hostess. He was severely reprimanded for inviting to dinner a local clergyman he met on his travels. He may well have visited the large and prosperous farms nearby, with their progressive tenants, and noted their potential as sporting country as well as their agricultural qualities. The solitariness of his visit was broken by a visit to Norwich for the assizes, where his father thought he should meet some of the local gentry. The outing provided welcome relief, and he particularly enjoyed the Assembly Ball, where he danced with Miss Pratt of Ryston Hall. On his return to Holkham he was duly quizzed by the old lady, who thought Miss Pratt was in no way good enough for him. 'Sir, you should have led out no one of lower rank than Miss Walpole.'[30] Finally, at the end of July 1771, he was able to leave the oppressive atmosphere of Holkham for his travels abroad.

This description of Coke's visit to his great aunt is drawn almost entirely from the two unfinished biographies in the Holkham archives. As they may well have been partly dictated by Coke himself, they give perhaps too harsh a view of Lady Margaret. Others wrote of her 'kindness of heart'. Mrs Poyntz of nearby South Creake wrote in 1766 of a visit from Lady Leicester bringing a gift of fish. Her behaviour was 'civil and pleasant', but perhaps a hint of her aloofness is implied in the words, 'she does great good, keeps many families, but is only loved by the poor as she will not mix with people here.' The hall was open on Tuesdays as well as Thursdays, which was 'her publick day and it is her delight to show the house'.[31] The number of visitors increased in the last years of the dowager countess's life, and a guide book of 1773 suggests that by then the open days had been extended. Holkham could be seen 'any day of the week, except Sunday, by noblemen and foreigners, but on Tuesdays only by other people'. While the author of the book admired the south front of the house with its portico and the views of it across the park against the background of trees, he was critical of the fact that there was no south entrance. 'This circumstance, after so fine an approach, and so long seeing the portico and expecting it to be the entrance, becomes a disappointment and a fault in the building.' The interior, however, was 'admirably adapted to the English way of living'. While some criticised the formal, and now unfashionable, straight drive and vistas, the author felt they were in keeping with the regular plan of the house.[32]

Whilst at Holkham Thomas would have had plenty of time to appreciate the importance of the Grand Tour to his great-uncle and the creation of Holkham Hall. Not only had he brought back many treasures in the form of the statues that stood in the statue gallery, the books that filled the library and the pictures that adorned the walls, but his travels and the architects he had met in Italy had been the inspiration for the building of Holkham itself. Thomas Coke had been a man more interested in things than in people and the project had been envisaged not so much as the creation of an ancestral home, but rather a 'temple to the arts'. In this spirit he created one of the most perfect Palladian mansion in Britain; a Classical villa worthy of the ancients 'that served to evoke Mount Parnassus, the haunt of Apollo, the god of poetry, and his companions, the Muses of Artistic Inspiration'.[33] 'Taste' was an all-pervading issue for the eighteenth-century gentleman, and as he had wandered through the lonely state rooms and galleries, young Thomas began to appreciate his predecessor's perfect understanding of the subject. The architecture, sculpture and paintings all embodied that taste, acquired and refined during his years of study in Italy. Seeing the treasures that his great-uncle had brought back created a background for his own travels abroad.

Foreign travel was an arduous business in the eighteenth century which first involved a Channel crossing, usually by the shortest route from Dover to Calais. This could take anything from six hours to several days. There could be problems with embarkation and landing if the winds were unfavourable. Lady Mary Coke was drenched in her attempt to reach France in 1764 and then could land no nearer than four miles from Calais.[34] When the Norfolk clergyman William Gunn set out with his wife and small daughter in 1792, he had a very rough crossing of over four hours to Boulogne. The mainsail split and his wife, daughter and servant were all ill.[35] From the coastal ports most travellers made their way to Paris, where they stayed for a while. The journey took the Gunns three days. Ann Gunn described it thus:

> Travelling post in France is not like England where you have good roads
> and horses, the latter we found very bad till within thirty miles of Paris.
> They don't go more than four or five miles an hour and at every Post they
> impose upon you in some way or other.[36]

Most then went on to Italy. The route was south to Lyons, a journey that would take five days, and then either by land through Savoy and over Mount Cenis to Turin, or to Nice, described by Ann Gunn as 'remarkable damp' with 'nothing to see'.[37] From Nice, travellers went along the coast, either taking a boat at Antibes for Genoa, or turned inland to cross the Alps at Tende. There was no route to Turin suited to carriages until the 1780s, so Coke would have had to go on horseback, and even this was hazardous and uncomfortable. However, he would have been rewarded by his first sight of real mountains. Even in the 1790s the route was difficult. Ann Gunn wrote that the road was never repaired, and quite impassable in the winter.[38] Following in his great-uncle's footsteps, Thomas first spent a few months at the university in Turin, described by Gunn as 'a new town and very beautifully laid out', where he met another young Norfolk gentleman, Martin Ffolkes Rishton, who was to become very much involved in promoting Coke's political career. Turin was famous for its military academy and riding school, where 'young gentlemen and strangers can be instructed at moderate expense'.[39] How much time Coke devoted to study in Turin is not clear, but he certainly made a name for himself as 'Le bel Anglais' at court, and acted as an escort to the King of Sardinia's daughter, the Princess of Savoy.[40]

While in Italy it did not seem to go against the scruples of someone brought up as a Whig patriot either to attend the courts of the various royal families who ruled despotically as absolute monarchs or to visit the Jacobite court in exile. Indeed, it is possible that alongside his Whiggism there was also a certain sympathy for the Jacobite cause.[41] In April 1772 he attended the marriage between the Young Pretender, Charles Edward Stuart, now 52 and an alcoholic, and the 19-year-old Princess Louise of Stolberg. He then accompanied them to Rome, where it was said that the young bride fell in love with Coke.[42] It would be no surprise if she preferred the company of this handsome Englishman of her own age to that of her drunken husband. It was rumoured that it was she who commissioned the magnificent portrait by Pompei Batoni which was finished in 1774 (Colour Plate 4). However, it is more likely that Coke commissioned it himself.[43] As well as the Italian artists, there were nearly 30 English painters living in Rome by the 1790s, at least five of whom painted portraits.[44] Nearly all young men returned from the Grand Tour with a portrait in their luggage.

In Rome there were many palaces, such as the Palazzo Doria Pamphilii and Jusiniani, which boasted works of art and would have welcomed visitors. Then there were numerous churches with their paintings and frescoes, museums and Roman remains to visit. Coke's Whig Protestantism did not prevent him seeking an audience with the Pope, Clement XIV, a man greatly respected by English travellers. William Gunn was received by the Pope along with three others when he travelled to Rome in 1785. 'His Holiness received us in a Morning Dress standing and convers'd with us very graciously on a Variety of Subjects for a quarter of an hour.'[45] From Rome, Coke visited Naples, Vesuvius and Herculaneum with his old school friend, Francis Rowdon.

When Coke left England there had been no firm plans for his return. The next logical stage in his career would be to enter parliament, and his father was anxious to secure him a Norfolk seat. The timing of his return therefore became dependent on the date of the election. By late 1773 plans were already being made, and letters from Norfolk were keeping Thomas informed of local affairs. Food riots in the autumn of that year were a topic of conversation between Coke and his friend Thomas Kerrich, who had relations living in Burnham Market near Holkham.[46] By November, Coke was in Florence, where he spent the winter amongst the numerous other English gentleman already there. Here he met for the first time Lady Mary Coke, the widow of his cousin Edward, who, if he had lived, would have inherited Holkham. Although an eccentric woman, she was always described as generous-spirited, and certainly her entry in her journal after meeting Thomas shows no trace of envy over the fact that he, rather than she and her husband, would be living at Holkham. 'He [Thomas] is a very pretty man ... as he is to have a very great estate, I am glad he is so worthy of it.'[47]

Letters from Thomas Kerrich to his sister in Burnham tell us something of Coke's last months abroad. In Florence, 'There are always a great many English here who live very sociably and agreeably together.'[48] In the morning they visited the Gallery, where they might stay for five or six hours looking at the pictures and meeting friends. A picture of the Tribuna of the Uffizi painted between 1772 and 1778 for Queen Charlotte by Johan Zoffany and now in the Royal Collection is obviously somewhat contrived, but gives an impression of the gallery at the time. The walls are filled with pictures from floor to ceiling and statues stand on the floor and on pedestals. The room is crowded with visitors, some chatting, some admiring the pictures, while other connoisseurs are studying the statues in minute detail.[49]

In the evening there were theatres and balls and dancing.[50] Together Coke and Kerrich visited Genoa and Milan as well as Tivoli Frascati, the site of Cicero's country house, and Hadrian's villa, where they were taken on a tour of what remained. They saw 'part of a temple in the Egyptian style and two theatres, one of which they say is ye most perfect now remaining ... I believe some of our party had quite enough of it; however we went to our lodgings and ate very heartily.'[51] Coke's great-uncle's taste for antiquities was exhibited in part by his young nephew, whose finest acquisition was a Roman mosaic representing a lion and leopard in combat which he inserted within one of William Kent's frames over the mantelpiece in the library at Holkham. Other mosaics from Hadrian's villa were inserted in two table tops in the saloon. In the spring of 1774 Coke began his journey home, visiting Vienna *en route*, where again he found great favour at court, particularly among the 'young beauties'.[52] There was no need for him to be home to canvass for his seat as there was unlikely to be any opposition, so he returned through Paris, where he again met Lady Mary, also on her way back to England. He arrived during the celebrations marking the accession of Louis XVI and Marie Antoinette to the throne – a time of optimism in France, where much was hoped for from the new king and queen. He was in England in time for the wedding of his younger sister, Elizabeth, to James Dutton of Sherborne (Gloucestershire) at Longford on 7 July 1774. He was never to travel abroad again.

So what had our young patriot gained from his travels? There was discussion at the time about the value of the Grand Tour to the many young men who embarked upon it. Some, such as Pope, were sceptical:

> Led by my hand, he sauntered Europe round,
> And gathered ev'ry Vice on Christian ground.[53]

Others found merit in 'seeing those countries which are famous as scenes of so much action in history',[54] and this was particularly true of the Classical world. But of more importance to the English country gentleman were the comparisons he was able to make with his own country as a result of travel. His visits to the many European courts would help him to appreciate the unique quality of 'British liberty'. In short, 'Every young man should go abroad to make him feel more attached to his own country.'[55] The Grand Tour could help to inculcate in young British gentlemen the idea that their loyalty was to Britain and its constitutional monarchy. Other countries were now seen as different and 'foreign'. Thus, patriotism was strengthened by travel overseas.

Coke had left England little more than a schoolboy and returned a confident young man with courtly manners. Foreign travel had given him a certain polish, which the rough and tumble of his country sports could not. Certainly he had been a social success with European royalty, and this ability to make friends and cut a popular figure was to remain with him all his life. However, he never made use of his European contacts, and showed little interest in diplomacy or foreign affairs, other than those of America, in his parliamentary career. He may have fraternised with Jacobites and met the Pope, but it seems that he returned with his Whig principles unblemished, to begin the next stage of his life. No letters or journals survive to reveal what he thought of the countries through which he travelled, but he did come back with a fine portrait and some Roman antiquities to add to the collection already at Holkham and to remind him of the continent he would never see again.

Chapter Two

EARLY PARLIAMENTARY CAREER, 1776–1783

Coke's parliamentary career got off to a false start. Parliament was dissolved in June 1774, and Wenman Coke was asked to stand for Norfolk, leaving the Derbyshire seat open for his son. Thomas had little enthusiasm for the contest, and when his opponent discovered that Thomas was still under 21 and therefore ineligible, he withdrew, while his father went on to represent Norfolk. When Wenman went to London in the autumn of 1774 for the start of the parliamentary session, Thomas went with him. The session brought fashionable society to town, including his sister Elizabeth and her husband, James Dutton, with James's youngest sister, Jane. The Dutton family had been established at Sherborne in Gloucestershire since 1551, and through the eighteenth century the family's wealth and lands accumulated. In 1776, on the death of his father, James inherited extensive estates throughout Gloucestershire, which included the Cotswold villages of Bibury, Aldsworth, Northleach and Windrush, as well as Standish, Moreton Valence and Stonehouse. The county parliamentary seat was controlled for the Whigs by the Berkeley family and for the Tories by the Duke of Beaufort. When James inherited, the Earl of Berkeley granted him the honorary titles of Deputy Constable of St Briavels and Deputy Keeper of the Forest of Dean. Between 1778 and 1779 he was High Sheriff. As one of the leading Whig families in Gloucestershire, Duttons were frequently Deputy Lord Lieutenant. When the parliamentary seat that the earl controlled became vacant in 1781 James became an MP, remaining in the Commons until 1784, when he was created first Baron Sherborne.

The marriage of Elizabeth Coke to James Dutton in 1774 allowed for close contacts between the two families. As members of the Whig elite, they had long known each other, and a marriage between Wenman Coke's daughter and the heir to a sizeable Gloucestershire estate was considered a good match. Sherborne house, overlooking the Windrush and on the site of a monastic grange, was on a large scale. From the outside it might have seemed old-fashioned, but the interior was much more up to date, having been remodelled by William Kent between 1723 and 1730.[1] Perhaps about the same time an impressive Classical stable block of Cotswold stone was built around two yards beside the house. The park too was extensive.

Traditionally deer parks were separate from the house and this was the case at Sherborne, but the eighteenth-century landscaping by Bridgeman had made the two visually one. As well as the Gloucestershire estate, the family owned land in Hampshire and Ireland, which in 1823 was bringing in an income of £25,000.[2] Thus the house and lands which Elizabeth and James took over in 1776 would have supported a lifestyle which, if somewhat less magnificent, was on a par with that which Coke inherited at Holkham in the same year.

In spite of the local standing of the Dutton family, when Thomas Coke suggested to his father that he might marry James's 20-year-old sister Jane, Wenham was far from impressed. It was his aim to marry his son to the heiress of a Derbyshire baronet who was expected to inherit the enormous fortune of £40,000 a year. However, she was said to be no beauty, and may indeed have been slightly deformed.[3] This difference over marriage meant that relations with his father grew more strained. There is no record of the dowry which Jane Dutton brought with her, but the marriage settlements for her two sisters were £10,000 each.[4] Thomas was spending more time in the Duttons' company, renting with James a house in Oxfordshire where they established a pack of hounds and were able to pursue their passion for fox hunting. In these pleasurable pursuits, the autumn of 1774 moved into the winter of 1774/5.[5]

In February 1775, the dowager Lady Leicester died, thus frustrating her determination to outlive her nephew, and Wenman Coke inherited Holkham. Young Thomas Coke was given the largely ceremonial title of Steward, Coroner and Bailiff of the Duchy of Lancaster in Norfolk and on 6 May he celebrated his twenty-first birthday. Now that he had reached the age of majority he was able to announce his intended marriage to Jane without his father's approval. Although Thomas stood to inherit a large estate, he still had no income of his own, and so, without his father's help, marriage was impractical. Thomas's friend and the owner of the neighbouring Norfolk estate of Gunton, Sir Harbord Harbord, was finally able to persuade Wenman to accept the inevitable and although he did not attend the wedding at Sherborne on 5 October 1775, he agreed to provide his son with an allowance.[6] Thus began what was by all accounts a happy marriage of 25 years, cut short by Jane's death at the age of 47 in 1800.

The winter of 1775/6 saw the newly-weds installed in the rambling old manor house of Godwick, near Tittleshall at the heart of the original Coke estates. Built at the end of the sixteenth century by the founder of the family wealth, Chief Justice Coke, it was not unlike, although on a smaller scale, the Dutton home at Sherborne. Thomas moved his hounds there and the winter was spent entertaining friends and hunting. His father, in contrast, spent little time in Norfolk, but was instead in London on parliamentary business. All this was set to change, when, in April 1776, Wenman Coke was taken seriously ill and Thomas was called London to be with him. The unlikely diagnosis was that Wenman's 'naturally robust constitution' had been destroyed by his sedentary habits. 'The ultimate consequence was a constipation which medicine could not remove',[7] and he died on 10 April at the age of 59, leaving Thomas the owner of Holkham and the natural choice to succeed to his father in parliament.

The Norfolk landed society of which the young Coke now found himself a member consisted of about 40 families owning more than 5,000 acres apiece. All lived in country houses often surrounded by parks, exercising political, economic and social control across their extensive estates. However, only a handful of these families involved themselves in the expensive business of national politics. Many of the largest estates were to be found in the north and west of the county, and so many of the influential Whig families lived within about 20 miles of each other (see map page 83). In the east, land ownership was much more widely distributed, with far fewer large houses within parks. By far the largest estate (about 30,000 acres) was that owned by Coke himself. The next two in size, covering between 15,000 and 18,000 acres and within 10 miles of Holkham, were those of the Townshend and Walpole families. Both had provided prominent political leaders of the previous generation. Robert Walpole of Houghton is recognised as the first man to hold a position comparable to that of a modern prime minister. From a relatively modest gentry family, Walpole was able to amass enough wealth by 1722 to employ Colen Campbell to design an impressive hall at Houghton. Campbell's *Vitruvius Britannicus*, the second part of which was published in 1717, was aimed at attracting the great Whig landowners to Classical architecture, and in it he dedicated a design to Robert Walpole. The interiors at Houghton were the work of the fashionable interior and landscape designer William Kent, and the result was a magnificent mansion in which Walpole placed his collection of pictures, reckoned to be the finest of its day. By Coke's day the hall had fallen into disrepair and the pictures sold off to Catharine the Great of Russia to pay the debts of Robert's grandson, George. On his death in 1791, there was no direct heir and the estate and the title of Earl of Orford was inherited by Robert's now elderly third son, the famous diarist and builder of Strawberry Hill, Horace Walpole. He never took up residence at Houghton, and died in 1797, when the estate passed to the Marquis of Cholmondesley, grandson of Robert's daughter, Mary. With estates elsewhere, he probably spent little time in Norfolk. Instead, the Walpole name and interests continued to be represented by the family of Robert's brother, Horatio, created Baron Walpole, who accumulated an estate and built a Classical mansion at nearby Wolterton. Although the family never again rose to the political heights achieved by Robert in the 1720s, members continued to represent a Norfolk seat. There were only five years out of the 150 before 1832 when a Walpole was not the Member of Parliament for Kings Lynn.

The neighbouring Townshend family had been of local standing longer then the Walpoles and had held prominent positions in Norfolk from the seventeenth century. Horatio Townshend's support for the restored monarchy was rewarded in

1682 with the title Viscount Townshend of Raynham. His fine brick mansion dates from the early seventeenth century and so belongs to an earlier school of architecture than Houghton. Although probably influenced by the work of Inigo Jones, it was built to the design of Roger Townshend, an amateur architect in his own right. Roger died in 1637, and work on the house was resumed by Horatio after the Restoration. William Kent worked here in the 1730s to create interiors of great dignity. Horatio's son, Charles Townshend, rose to be one of the two Secretaries of State alongside Robert Walpole in the 1720s, until he was finally outmanoeuvred by Walpole and resigned his position, leaving Walpole with sole power, in 1730. The family continued to represent Great Yarmouth, holding a seat there almost continuously throughout the eighteenth century. While both families were seen as Whig grandees, they played little part in the great issues of the period under consideration in this book.

Coke's closest friend and political ally in the early days was William Windham of Felbrigg in north-east Norfolk. A man of great intellect and friend of Samuel Johnson, his interests were academic rather than political. His diary is full of the self-doubt which dominated both his intellectual and political career. He often expressed a wistful regret that he had not stuck to mathematics, about which he wrote copious notes, rather than entering politics. 'What a cause of regret, that in years and years I should have proceeded so little hitherto! Why might not I, at this moment, have been among the first rate of mathematicians.'[8] For many years Windham held one of the Norwich seats. He and Coke had political differences over the conduct of the war against Napoleon, but remained close friends. The family was a long-established one, and in 1608 the dilapidated old manor house of Felbrigg was pulled down. Work on a new mansion began in 1615 and was finished five years later. A Classical wing was added in the 1670s. William Windham's father, also a William, on returning from the Grand Tour in 1750 remodelled the interior, creating a great Gothic library to the plans of James Paine, building a new service wing and redesigning the existing state rooms in the west wing. His son inherited his father's passion for books, which led to the blocking of the great bay window in the library with bookcases matching Paine's Gothic ones to house his ever-increasing collection. This included books given him by Samuel Johnson in the old man's final days as a mark of recognition of Windham's devotion to learning. It was his scholarly interests that ensured his friendship with Coke would survive their political differences.

Another Whig family with whom Coke was closely associated were the Astleys of Melton Constable. The building of Melton Constable Hall within its park was begun in 1665 and it is described as 'probably one of the six most important houses of its date in England'.[9] Built in red brick with stone dressings, this Restoration house is nearly square, with a flight of steps up to a front door to the south and a pedimented porch to the west. The fine plasterwork ceilings added to the elegance of the Astley home, and Sir Jacob was a Norfolk MPs at much the same time as Coke. Other prominent Whig families with whom Coke was closely associated included the Hobarts of Blickling in their Jacobean mansion; the Harbords of Gunton, whose hall was designed by Mathew Brettingham (one of the architects of Holkham); and Lord Albemarle (of the Keppel family), who lived in the south of the county, at Quidenham, a much-altered early eighteenth-century house not bought by Lord Albemarle until 1762. In the late eighteenth century the family bought from the Wodehouses the smaller house at East Lexham, neighbouring the Holkham mid-Norfolk estate.

Coke's main Tory rivals were the Wodehouse family, created Earls of Kimberley in 1866, who owned a slightly smaller estate about 25 miles to the south-east of Holkham. Kimberley Hall was a fine red-brick eighteenth-century mansion surrounded by an impressive park. Sir John Wodehouse and Coke were both county MPs from 1790 to 1797, when Sir John was ennobled as Lord Wodehouse and moved to the House of Lords. This handful of landowners, all owning over 8,000 acres, with great houses, some dating back to the early seventeenth century, others new Classical creations, were in close touch with each other, employing the same architects, agricultural advisers, landscape gardeners and interior designers. In London their portraits were painted by the same artists and busts executed by the same sculptors. It was these men, living in seclusion within their wooded, landscaped parks, who also controlled the politics of the county.

Candidates were thus drawn from a very small elite group, and fighting an election could be an expensive business. For this reason contested elections were kept to a minimum, with the rival candidates coming to agreements between themselves before election day. Between 1790 and 1815 only three out of ten elections were contested. Coke was frequently accused of using his great wealth too freely or 'shaking his purse at the county', and certainly the contested election of 1806 cost Coke and Windham £32,000 and Wodehouse £20,000.

Not only did candidates need money before they put themselves forward, but also 'county standing', best attained after long years of owning a family estate. The County Members for Norfolk between 1750 and 1820 were provided by the Townshends of Raynham, the Wodehouses of Kimberley, the De Greys of Merton, the Astleys of Melton Constable, and the Windhams of Felbrigg, as well as the Cokes of Holkham. The Windhams were not a wealthy family, but their 400 years of continuous occupation of their estate gave them the necessary position in society.

With the great expense that elections could involve and the need to spend time in London, especially during the shooting season, the question arises as to why there was such competition for seats in parliament. Firstly there was no doubt a genuine desire to serve and influence affairs. However defective the system of electing, those who became MPs felt a genuine pride in their position which brought prestige to the family. Coke revelled in his independence as a County Member owing his seat to nothing other than his own county influence and standing. When refusing a title he took pride in the fact that kings might make peers, but not County Members. In his 1790 address to the electors he said, 'The situation of being the representative of the freeholders of Norfolk is certainly the proudest that an independent man can be placed in.'[10] The honour of being elected was far more attractive than having to purchase a rotten borough. Of the other 10 Norfolk seats, only Norwich, with its fierce rivalry between the two main political groupings and its 3,000 voters, was relatively free from outside interference. In contrast, Great Yarmouth contained 800 voters in 1790 rising to 1,400 by 1818, and the Townshend family (with the exception of 1796, when an independent party made up largely of Dissenters managed to oust their candidate) controlled elections after 1790 through their patronage of the Corporation. Kings Lynn's electorate was much smaller, at about 300, and was loosely controlled by the Walpoles, the Turners, a leading merchant family, and the Ffolkes of nearby Hillington. Thetford, with only 30 voters was variously in the hands of the Duke of Grafton, the Duke of Norfolk and Lord Petre, while Castle Rising, a truly 'rotten borough' was dominated by the interests of the Walpoles and Howards of Castle Rising.[11] From 1802 to 1819 the seat at Thetford was held by the Foxite Whig Thomas Creevey, who probably never visited the town. While arguing vociferously for the abolition of rotten boroughs, he was happy to represent one. In the contested election of 1806 when Coke and Windham stood against Wodehouse, Windham made it clear that he would regard it a greater honour to represent Norfolk than the rotten borough which he had been offered. However, when the election was declared null and void, he fell back on the offer of the seat for New Romney in Hampshire.

Almost immediately after Wenman's death, Sir Harbord Harbord and the remaining Norfolk MP, Sir Edward Astley of Melton Constable, visited Thomas at his house in Grosvenor Square to ask him to stand in place of his father. In later years he claimed that it was with great reluctance that he agreed. He did not see himself as an orator or a politician. At only 21, he had hoped to enjoy the freedom of his newly inherited estates, with the opportunities for sport they provided, for a few more years before taking on public responsibilities. But when his visitors pointed out that if he did not stand, a Tory might take the seat, 'my blood chilled all over me from my head to my foot, and I came forward'. Educated, as he had been 'that a Tory was no friend to liberty or the [1688] revolution' but instead a 'supporter of bribery and corruptness and of all the evils of oligarchy'[12], his duty seemed clear. Two days after his father's death, on 12 April 1776, Thomas issued his manifesto from his London address. In it he promised the electorate to give 'the most universal attention to your particular interests, and to the honour, liberty and essential well-being of my country'.[13] He then travelled back to Norfolk to take possession of his inheritance and to organise his election. Whilst there is little doubt that Coke saw it as his patriotic duty to stand, a seat in the Commons could be extremely beneficial to landowners. Much late eighteenth-century legislation involved turnpike, canal and enclosure bills, and it was the county and borough members who saw them through the Commons and made sure their friends and family connections in the House of Lords turned the bills into Acts. Enclosure Acts were of particular concern to Coke. Now the largest landowner in Norfolk, but with little other source of wealth, he was acutely interested in maximising the income from his estates.

In the event, there was no opposition to Coke's election. He was unanimously nominated at a meeting in the Shire Hall, Norwich, on 27 April. This was followed by a further manifesto asking for the support of the freeholders of the county and promising to represent not only 'the real Interests of this County, but likewise to the Honour, Liberty and Welfare of the Nation'.[14]

The election took place in Norwich on 8 May in a festive atmosphere, much in the style depicted in Hogarth's paintings of a generation before. Coke rode at the head of his electorate, who included all those freeholders who could be persuaded to come by the offer of free transport and hospitality in Norwich over the days of the election. They entered Norwich from the north, through open countryside which only gave way to market gardens and orchards as they approached the city, still at this time mostly contained within its flint walls. Entering through the wide arch under the castellated tower of St Augustine's gate, one of the eleven city gates still standing, they rode through the open gardens of the outer city before entering the narrow streets and bustle of a busy centre. The city walls enclosed an area large enough to contain a fast-growing population and still leave room for large houses and gardens at the outer edge. Many well-to-do citizens still lived within the walls in imposing brick mansions, and the only slightly less well off in elegant terraces such as that in Surrey Street designed by the fashionable local architect–builder, Thomas Ivory. The city's wealth derived both from weaving on the 6,000 looms still producing cloth (which supported dyers, combers and hotpressers)[15] and from its importance as a commercial and financial centre for a prosperous hinterland. Politically the city was divided, but a powerful Whig group was led by the wealthy Quaker banker John Gurney, who still lived within the city walls in Magdelan Street. It was also an important cultural and social centre, with an influential Dissenting community centred on the Unitarian and Independent chapels in Colegate. One of the most important open places was the market-place, occupying a large, rectangular piece of flat ground below the castle, with the church of St Peter Mancroft on one side and the fine medieval guild-hall, with its chequer-board flint front, on the other. Every day market women came here with their stalls. It was also the venue for all public parades, bonfires and celebrations, and it was where the polling booths were put up for city and county elections. Alongside the market was Gentleman's Walk, which 'on a market day is thronged with a collection of very interesting characters; the merchant, the manufacturer, the magistrate, the provincial yeoman, the affluent landlord, the thrifty and thriving tenant, the independent farmer, the clergy, faculty, barristers and all the various characters of polished and professional society'.[16] Into this busy and populous throng rode Coke and his supporters.

Voting took place over several days and was in public. Coke was expected to provide refreshment before, during and after the hustings. At a later election in 1796, the agent in Swaffham, William Stuckey, engaged 'all the available chaises, carts and horses' in the area to prevent them being taken by the opposition.[17] The estate tenants were also expected to help with transport of voters. When the election was a contested one they were instructed to 'bring up all the wavering and uncertain voters as early in the day as possible and never lose sight of them from the time of leaving their homes until they poll ... the staunch voters should follow the uncertain ones. It will be desirable that as many of Mr Coke's principal tenants and other friends should attend the nominations as can do so conveniently. ... You will please use every honourable means to procure voters: <u>perseverance generally gains the battle</u> and do not forget that both your and your worthy landlord's honour is at stake in this contest.'[18] Thomas Moore, the tenant of Hall Farm, Warham, went to Norwich to support his landlord at the 1802 election and on 12 July 'attended the poll booths the chief part of the day after riding in the cavalcade to meet Mr Coke and Sir Jacob Astley'. At the next election, in 1807, he rode in the cavalcade 'on a borrowed horse'.[19]

Once the election was over, the proceedings continued in style. At the 1784 election, Coke enlisted 39 chairmen at a guinea each to carry the ornate electioneering chair, now in Strangers' Hall Museum, Norwich. Two marshals, also at a guinea, and nearly 300 stavesmen were paid between 5s and 10s 6d to walk alongside him. Cockades were provided for the impressive procession around the city.[20] On 8 May the day began with speeches. Coke remembered, some 50 years later, that he had declared to the freeholders that he was 'a Whig of the old school, a lover of the principles of the Revolution of 1688 and by that line of conduct my course shall be governed'.[21] The day ended with his being chaired around the market place through the by now raucous crowd. The successful candidate was expected to pay for the church bells to be rung and for music to accompany him on his triumphal route.

The election of 1776 was necessitated by the death of an MP, but even at the time of general elections, most seats were uncontested. Nationally, only about one in six contests went to a poll, because of the great expense involved. Coke was said to have spent £500,000 on elections during his long parliamentary career, but surely this is an exaggeration.[22] Polling could take up to 15 days, although usually it only went on for a week. In the contested elections of 1802, Coke had only polled 32 votes in the first few days, with seven days still left. In the event the two Whig candidates, Jacob Astley of Melton Constable and Coke, were returned after a scrutiny of the poll because the contest was so close. Generally, unlike the situation in many smaller boroughs, there were enough influences at work in counties for their members to be relatively independent of aristocratic pressures. This was particularly true of a large and populous county such as Norfolk. The budding landscape designer Humphrey Repton had successfully managed Windham's campaign for a Norwich seat in 1784 and in 1788 he performed the same function for Coke. Although the great land-owners held a practical monopoly of political activity, there was a potential for independence amongst the small but homogeneous group of lesser players. Their support had to be constantly nurtured by those who sought their votes.

Before the electoral reforms of 1832, the Norfolk electorate of 40s (£2.00) free-holders, each with two votes (one for each of the two MPs returned for the county), was made up of about 5,000 men, or about an eighth of the adult male population, and included clergy, gentry, farmers and some craftsmen.[23] Even though freeholding gave voters some independence, the fact that voting was at the public hustings meant there could be no secret over who supported whom. Many of the tenant farmers on the Holkham estate and elsewhere were also freeholders of small estates and could be expected to vote as their landlords wished. Tradesmen, too, depended on good relationships with the gentry and would vote to please their customers. In 1784, Coke had hoped to win over the voters of Walsingham, but because he did little business in the town, he could attract little support. The major family, the Hills', 'connections in trade were so great as to influence half the town'.[24] Two other less tangible influences were 'gratitude' and 'obligation', which could be manipulated by the candidates. It was local and family interests rather than party loyalties that largely determined how the freeholders voted. This was particularly true at a time when the parties themselves were fluid and it was small groupings within the Whig party which dominated politics.

The House of Commons Coke entered was very different from that of today. It met in St Stephen's Chapel, remodelled by Sir Christopher Wren. The total seating capacity was under 500, but rarely was anything like the full complement of 658 MPs present and sometimes the quorum of 40 could not be reached. Speaking was generally left to a small handful of activists, which did not normally include Coke. In 1790 57 per cent of speeches were made by only 19 speakers, all of whom spoke more than a hundred times, many of them for far too long.[25]

Although Coke was only 22 when he was first elected and thus the youngest member at the time, he would not have felt out of place in the House, where about 15 per cent of members were under 30.[26] At least one fifth of the members of each parliament between 1754 and 1790 had been to either Westminster School or Eton, with those from Eton increasing in influence by the end of the period.[27] Nearly 170 were heirs or younger sons of peers and nearly all were landowners of substance. The ideal MP was the 'country gentleman': self-sufficient, independent of court and party alike, who judged every issue in the House on its merits.[28] Coke certainly fitted into this category. The turnover of MPs was low, with over half sitting for more than 10 years. By the time Coke retired from the House in 1832 he had been a member for 50 years and had thus earned the title 'Father of the House'.

We know little of Coke's early days in parliament. His election in May 1776 was just before the session ended on 26 May, but he appears to have been busy in London in the early summer. Very soon after his arrival he was invited to dinner with the leader of the main opposition party, Lord Rockingham, where he was introduced to the leading Whig, Charles James Fox.[29] This was the beginning of what was to prove a lifelong friendship. The two men were very different. Fox was renowned for his flamboyant lifestyle. His chaotic private life and wild gambling which resulted in his amassing huge debts were legendry, even in an age when such things were not unusual. People lent him money knowing that there was very little prospect that it would ever be returned. They did so out of friendship or possibly as an investment in a promising career. 'Their generosity fulfilled the prophesy that Fox would live under a divine dispensation from the rules that governed other men.'[30] Coke on the other hand seems to have had an exemplary domestic life and was no gambler. He was one of a group who, in 1793 set about fund-raising to help clear Fox's debts yet again.[31] Giving money at a time when Fox's views on the French Revolution were becoming well known could be seen as a political act and a staggering £61,000 was subscribed, with Coke amongst the largest subscribers. However, even this large sum did not provide long-term financial security, and the last years of Fox's life were clouded by the realisation that he was unable to provide his wife and children with a secure future. He died with debts of £10,000.[32] Fox was also known for his ability as an orator. He was able to speak for up to five hours at a time, and was said to keep his audience transfixed. Coke on the other hand was, at his own admission, not a great speaker and contributed infrequently to debates.

However, Fox and Coke shared a passionate political belief in 'liberty'. This rather vague term included support of the right to religious freedom both for Roman Catholics and Dissenters. Although not a religious man, Fox proclaimed in 1791 that 'toleration in religion was one of the first great rights of man',[33] and Coke certainly voted with him on this issue. Fox was also a fierce opponent of the slave trade and a supporter of William Wilberforce. His support for parliamentary reform was, however, more limited. His belief that voting should be linked to property was shared by the other great Whig landowners. He thus resisted moves towards universal suffrage, while supporting the need to enfranchise more property owners. He was more concerned with the corruption from above than the views of the unrepresented below.

Fox and Coke also shared an equally passionate obsession with field sports, and Fox was frequently to be found at Holkham enjoying the shooting. But there was another side to this complex and charismatic man. He had a deep interest in the classics, literature and art. A surviving letter written from Holkham is to his nephew,

Lord Holland, who was on the Grand Tour, discusses at length the merits of various great painters. [34] There is no doubt that as well as the sport, he would have enjoyed many hours in the library and amongst the pictures in the hall. In later years Coke remembered their first meeting: 'When I first went into Parliament I attached myself to Fox and clung to him through life. I lived in the closest bond of friendship with him. He was a friend of the people, the practiser of every kindness and generosity, the advocate of civil and religious liberty.'[35] That this friendship was seen as dangerous by the government is suggested by Coke's own assertion, later in life, that he was offered a peerage in an attempt to win him over to Lord North's rival camp within two months of entering parliament. This he vehemently declined, reasserting his support of Fox.[36] There is only Coke's word for this, but if true, it suggests that the North government feared his influence and that he was expected to play a prominent part in affairs.

Coke entered the Commons in 1776, at a crucial moment in both home and foreign affairs. Lord North was at the head of the administration and still had a sizeable majority in the House of Commons. North was a politician of considerable ability who had led the government since 1770. His economic competence meant it had been a period of stability and vigorous economic recovery, which had resulted in the collapse of unity amongst the opposition groups. These split into two, under the leaders of the two previous administrations, Lord Rockingham, who had led a short-lived government from 1765–6, and was now supported by Fox and his group, and Pitt the Elder, now Lord Chatham, who had returned to power to lead an administration between 1766 and 1768, and now led a second, more radical group. Most importantly, however, Lord North had the support of the king and the court.

This period of steady growth and political calm was brought to an abrupt end by the government's mishandling of affairs with the American colonies, and North was coming under increasing attack for his American policy from Fox and the grouping of Whigs around Rockingham. Following the controversial and disastrous attempt to impose taxes on the Americans in the form of import duties, and the climb-down resulting in the removal of all taxes except those on tea, Rockingham had been responsible for the hated Declaratory Act of 1766 which reaffirmed the supremacy of parliament and its right to legislate for the colonies. By 1770 North was in power and relations had moved from bad to worse. This culminated in 1773 with the throwing of a cargo of tea into the harbour at Boston by protesters. The 13 states held a congress in Philadelphia in 1774 and the first shots were fired in the American War of Independence at Lexington in 1775. Everything that Fox and the Rockingham Whigs had feared had come to pass. However, as they had been responsible when in power for the early mishandling of the situation, they were in a difficult position and concentrated their criticism on the confrontational approach of the government and the inefficiency and corruptness of the management of the war.

As an independent member for a county constituency, Coke stood by the principles of the Glorious Revolution of 1688 and the subsequent Bill of Rights which was seen to have ended the arbitrary power of the king and replaced it with a free and independent parliament. From this flowed a belief in justice and fair play, an unbroken attachment to civil and religious freedom and hatred of all intolerance, oppression and coercion. The support of these principles, both at home and abroad, was the duty of every patriot, and it was clear to Coke that many of these principles had been broken in the North government's handling of American affairs. Patriotism and support of the colonists were thus entirely compatible.

The cause of the colonists was one that united the opposition against Lord North. When parliament reopened on 31 October 1776, the king announced in his speech that more troops would have to be sent to America to put down the rebellion. The formal address of thanks from the Commons was proposed, but an amendment was then moved deprecating the attitude of the British government towards the colonies. Coke, along with Fox and the radical John Wilkes, was amongst those supporting the amendment.[37]

We hear little more of Coke's public activities until January 1778. By this time the war in America had gone from bad to worse, and after the defeat at Saratoga in October 1777 it was clear that any chance of a British victory would involve a long and expensive war. Even George III appreciated the likely consequences of the situation and is said to have commented optimistically that things were 'serious, but not without remedy'.[38] With feelings against the conflict rising both in parliament and the country, Lord North could see that his support was weakening and asked the king to accept his resignation. This he refused. North was then plunged into a depression which led to indecision and inactivity. The king was finding it more and more difficult to raise funds for the war and resorted to attempting to raise voluntary donations from his 'loyal' subjects. A meeting was held in January 1778 at the Maid's Head opposite the cathedral in Norwich to 'consider the best means to assist the exertions of the British Empire in support of its constitutional authority'. It was chaired by Sir John Wodehouse, a supporter of North and opponent of Fox. In less than an hour £4,500 was subscribed.[39] The meeting was also attended by Thomas Coke and other opponents of North, including William Windham of Felbrigg, who made an impassioned speech in support of the American colonies, a copy of which survives in the Holkham archives.[40]

Citing the government's mismanagement and misrepresentation of the war, Windham pointed to a campaign which had resulted only in 'disappointment, shame and dishonour.' The sums of money granted by parliament had been squandered and now the 'Ministry had the effrontery to apply for voluntary contributions', claiming that the country was in danger of foreign invasion as a result of the support the French were giving the Americans. This, Windham claimed, had all been predicted by the opposition, and the giving of donations would serve only to keep the current

Plate 1: Coke's friend and sometime political colleague, William Windham (1750–1810) in a sketch by Humphrey Repton. He is shown delivering his first political speech at a meeting in Norwich attended by both men and called as a protest against the American war. He condemned the war as one which was bringing 'disappointment, shame and dishonour' to Britain. A copy of the speech exists amongst the Holkham archives.

government in power. Indeed the government's removal would, in fact, be a step towards the end 'we are all interested in, peace and reconciliation with America'. Windham, Coke and their supporters then withdrew to the nearby Swan public house, where they began considering the preparation of a petition to the king from 'the Nobility, Gentry, Clergy, Freeholders and Inhabitants of the County of Norfolk' requesting the ending of the war, two copies of which survive in the Holkham archives. The very existence of these hand-written copies at Holkham is a clear indication of Coke's support for, and involvement in, the anti-war campaign.[41] This was Windham's first public speech and Humphrey Repton sketched him holding the Norfolk petition. When it was finally presented to parliament on 17 February 1778, probably by Coke,[42] it had been signed by 5,400 Norfolk people. Although blame for the disasters of the war was laid firmly at the feet of the government rather than the king, it was taken as a personal attack by George III, who regarded Coke with animosity for the rest of his life. Until the American war national politics had hardly touched the local squires, and although they left the speaking at the meeting to peers and MPs, the number of signatures suggests that things were changing. The Norfolk petition predated by a year the main period of the petitioning movement led by the Yorkshire clergyman Christopher Wyvill, with its different concerns.

Wyvill called a Yorkshire County Meeting in December 1779 to protest against high wartime expenditure and the levels of taxation. Yorkshire, like Norwich, was suffering from a downturn in the textile trade as a result of the war. In the early months of 1780 petitions from a further 26 counties and a dozen boroughs were presented. The main supporters of the movement in the countryside were independent-minded landowners and tenants, while in the towns it attracted radical professionals, shopkeepers and artisans. Unlike the Norfolk petition, which called only for the dismissal of the king's ministers who had 'been entrusted with an almost unbounded Command over the Public Purse, and the consequence has been an almost unbroken series of Calamity and Disgrace', those inspired by Wyvill called for wider reforms. These included triennial parliaments, the abolition of rotten boroughs, the restoration of the independence of the House of Commons from court influence and a cautious reform of the electoral system. It was clear that there were those in the counties who were increasingly taking an independent line in political matters.[43]

In the same month that the Norfolk petition was presented, North put forward peace proposals which involved the repeal of the Declaratory Act and the appointment of commissioners to discuss peace. While this might have satisfied the Americans before the war, now nothing short of full independence was acceptable. At the same time, Fox, the most prominent of the Rockingham Whigs, proposed a motion that no more troops should be sent to America, for which he won 165 votes. Meanwhile the French had joined the war on the side of the colonists, and in June 1779 Spain also allied herself against Britain.

As hostilities dragged on, both the war and North became more and more unpopular, and the attacks of the opposition on his handling of the war increased. Rockingham and Shelburne, building on the public support aroused by the petitioning movement, argued for American independence, although they hoped that economic ties might remain. The surrender of General Cornwallis at Yorktown in October 1781 marked the end of credibility for North's ministry. On 22 February 1782 the MP for Bury St Edmunds in Suffolk, General Conway, put the motion 'That an humble address be presented to His Majesty, that, taking into his royal consideration the many and great calamities which have attended the present unfortunate war, and the heavy burdens on his loyal and affectionate people, he will be pleased graciously to listen to the humble prayer and advice of his faithful Commons, that the war on the continent of North America may no longer be pursued for the impractical purpose of reducing the inhabitants of that country to obedience by force.' The motion was seconded by Lord Cavendish.[44] Speakers from both sides followed and finally, at two in the morning, a vote was taken. General Conway's motion was defeated by one vote (194 to 193). Coke did not enter into the debate, but he may well have been there, and would undoubtedly have voted with the opposition.

However, other issues nearer home were on his mind. The preservation of game was becoming increasingly an issue in the late eighteenth century, and from 1784 the government was able to cash in on this developing gentlemanly sport by charging game licence duties on owners of shoots. Improved guns coupled with the enclosure of commons, the growing of turnips as a field crop and increased acreages of grain provided a farming environment particularly hospitable to pheasants and partridges. The preservation of game was increasingly possible within compact, ring-fenced estates. The result was a series of laws establishing the landowner's right to game on his land with ever harsher punishments for those who attempted to poach. Two days after the defeat of General Conway's motion, on 25 February, and in the midst of heightened excitement over the future of the American colonies, Coke spoke on the subject of the game laws. As an enthusiastic sportsman Coke had a considerable personal interest in the subject, but even he thought they should be reviewed and relaxed. 'Combinations had been formed in the County against the execution of these laws and some lives had been lost.' A surprising ally was the Tory-inclined Charles Turner, an MP for Kings Lynn, who described the laws as 'cruel and oppressive on the poor'. 'It was a shame that the House should always be enacting laws for the safety of gentlemen; he wished they would make some for the good of the poor: if gentlemen were not safe in their homes it was because the poor were oppressed.' He went on to make the rather unlikely statement that 'if he had been a common man he would have been a poacher in spite of all the laws'. The laws needed revising and 'stripping of half their severity'. Sir Edward Astley spoke of 'poachers armed with firearms' and 'the necessity of revising the laws',[45] but it was not until 1827 that the use of spring-guns and mantraps was forbidden. These were issues dear to Coke's heart, as well as of concern to his electors, and were ground on which he felt far more secure than complex foreign policy. However, the problem of America would not go away and General Conway announced that he would be reintroducing his motion, with slightly different wording, on the following day. His reason was that the government's majority had been only one, and many members had been absent from the House (355 had voted out of a possible total of 658).

A new motion was duly put forward at 4.30 on the afternoon of 27 February, but Coke was not one of the many speakers. A vote was finally taken, and this time the number in favour was 234 and that against 215. Sir Edward Astley and Coke along with the MPs for Thetford and Norwich voted with the opposition, while the members for Yarmouth and the small borough of Castle Rising, whose two seats were controlled by North's supporter, George Walpole, the third Earl of Orford, voted with the government. Conway then moved that 'An humble address be presented to His Majesty' and that 'such members of this house as are His Majesty's most humble Privy Council, do humbly know his majesty's pleasure when he will be attended by this House'. The reply came back via Lord Hinchinbrook that the king would see them on 3 March at 3 p.m. at St James's Palace.[46]

There then follows perhaps the occasion for the most important and symbolic act of Coke's political career. As a Knight of the Shire, he had the right to attend court 'in his boots' rather than in formal court dress. As a mark of the esteem in which he held the position of a County Member and the little respect he had for the court, this young man of 28 chose to take the address to George III in his leather breeches, boots and spurs. The patriot stood face to face with the king. The drama of the occasion appealed to Coke, who then had his portrait painted by Gainsborough in the dress he had worn. Coke's biographer, Mrs Stirling, detected a look of pride and disdain in the face, indicating his attitude towards a king who rode roughshod over his colonial subjects.[47] To push the point home, he then hung the portrait in the most prominent position in the grand saloon at Holkham to be seen and admired by his many American friends and visitors (Colour Plate 6).

It's a good story, but did it ever happen like this? We only have Coke's word for it, in a much-quoted speech made at the 1821 sheep shearing. 'I carried up the Address as an English Country Gentleman in my leather breeches, boots and spurs.'[48] Is he simply creating a myth to demonstrate, metaphorically, his understanding of patriotism in the confrontation between arbitrary kingship and parliament, and particularly the independent County Member, or was the memory of an old man playing tricks nearly 40 years after the event? Certainly the fact that Coke voted for the motion is recorded, but *Parliamentary Register* describes the whole incident rather differently. On 4 March, the Speaker reported that 'The House' had attended the king with their address.[49] Coke was most probably there, but he was not the only Member of Parliament present. Indeed the *Norwich Mercury* recorded in its obituary that he went 'accompanied by Sir Fletcher Norton, the speaker and other members of the house'.[50] Maybe, in fact, Coke's statement in his speech was intentionally ambiguous, and he intended it to be interpreted in such a way as to suggest that he went to the king alone without exactly saying so. However, his statement that Fox had urged him to move that the address be carried to the throne is not supported by the report in *Parliamentary Register* and cannot be corroborated elsewhere.

The king was finally forced to accept the inevitable, and peace negotiations with the American colonies began. The king also had to accept the resignation of North, and in April 1782 a new government under Rockingham, with Shelburne and Fox as Secretaries of State, was formed. The disagreements between Fox and Shelburne soon became obvious. Fox wanted the colonies to be granted unconditional independence, but Shelburne felt that any agreement should be dependent on negotiations with America's European allies, France and Spain, against whom fighting was continuing. Fox and Shelburne were now bitter political enemies. Whilst it is possible that Rockingham would have been able to hold the disparate opposition together in government, this was not to be, as he died after a short illness on 1 July. The king then promoted Lord Shelburne First Lord of the Treasury (Prime Minister) and brought in the 22-year-old son of Lord Chatham, William Pitt, as Chancellor of the Exchequer. At this the rest of the Rockingham government resigned in protest and the Duke of Portland replaced Rockingham as leader. The king, they argued was taking on arbitrary powers and it should have been up to the cabinet to choose their own leader. They saw in the king's action the secret influence of court favouritism. Fox wrote to Coke acknowledging the blow to the strength and unity of the party caused by Rockingham's death and requesting his return to London, 'to assist in forming some plan for acting together in a body upon the same system and principles upon which we have hitherto acted'.[51] There followed a period of political chaos, and the government had to rely on the 200 independents for support. It was soon clear that Fox had been right and that nothing short of full independence for the American colonies was possible. The preliminaries to peace were agreed in November 1782, but when Shelburne presented his peace terms to the House of Commons in February 1783, they were narrowly defeated and at this a disheartened Shelburne handed in his resignation, followed by that of Fox. This left the king with very little choice but to negotiate with Fox and Portland, who felt themselves in a strong position. They demanded that Portland should be appointed to the treasury, and be free to choose his cabinet without royal interference. The king refused to countenance Fox and his supporters in the government. They were now actively courting the friendship of the king's estranged son and heir, George, Prince of Wales. For six weeks the country was without a government. The king now turned back to North, and in order to keep Shelburne out of office the unlikely Fox–North coalition was agreed in March 1783.

After months of uncertainty, a coalition government was finally formed in April 1783 with North and Fox as Secretaries of State and the Duke of Portland at its head. Coke was shocked by this arrangement, which he described as a 'revolting compact'. Fox presented it to him early one morning when he burst into his room and seated himself on the end of the bed to explain his great plan and to emphasise that it was North who would have to moderate his views and not himself.[52] Whilst Fox and North did succeed in ousting Shelburne, their collaboration was doomed to failure. The king reluctantly agreed to accept Fox in a governmental position. Parliament was not due to meet until November, and Fox, realising the danger of his position, was anxious that all his supporters should be there. He wrote to Coke, who was enjoying some autumn shooting at Holkham, urging him to come to London for the opening of parliament. Because he was nervous of the reaction he might get from his friend he ended his letter 'I assure you that it is for the support of the <u>cause</u> and not of any particular situation that I am thus earnest in requesting your attendance.'[53] In fact, the coalition government fell over the terms of the East India Bill, which seemed to confirm suspicions that Fox and North had only come together to share the spoils of office, and their government lasted only until December. Introduced into parliament a week after the session had been opened, the bill proposed that the government of India should be subject to a board of seven commissioners who would run the administration and control patronage.

Controversially, these commissioners would be appointed by the government for a term of four years rather than by the crown, thus challenging what the king saw as his constitutional right. Not surprisingly, the existing directors of the East India Company were also enraged and used their power to get the bill overturned by the House of Lords. On 16 December Fox again wrote to Coke pressing him to abandon the pheasants and return to London. 'We are beat in the Lords by the direct interference of the Court and if some vigorous measures are not immediately taken, the Parliament will be dissolved, and a system of influence established by acquiescence of the most dangerous kind yet attempted. I wish you at the same time to let this be known to any members you may happen to see, who have a spark of Whig principles left.'[54]

While the king and the opposition saw the bill as a method by which the government would be provided with a vastly increased scope for patronage to reward their friends and bribe their foes, Fox had in reality intended that it would at least partly cut down on the use of privilege and extend the range of responsible government. George III, who hated Fox, was determined to use this crisis to overturn the coalition, and asked the young Pitt to form a government in spite of the fact that Fox and North still commanded a majority in the House of Commons. On 19 December the ousting of Fox and North and their replacement by Pitt was announced to a packed House of Commons. When the Commons went into recess for Christmas, Fox was confident that Pitt would be easily overthrown and the king forced to accept the will of the Commons early in the following year.

Coke meanwhile, took no part in these debates and was dividing his time between Holkham and London. Many of the letters from Fox to Coke between 1776 and the 1790s are requests that he will come to London to support various bills. In an undated letter he asked Coke's support for several motions relating to Ireland that he intended to bring forward, as well as one on the general state of defence: 'therefore if you can, attend and get others to attend from the 15th for three weeks. Do you or any other Norfolk man mean to bring on the Corn bill or malt duty; we can make a far better muster of our old friends than I had thought.' He asked Coke if Sir Edward and Sir Jacob (Astley) or Sir Martin Ffolks, an MP for Kings Lynn, could be persuaded to come.[55] From time to time Fox could be tempted away from London for the shooting, and the game books at Holkham show him there for a few days in November most years in the 1790s. In December 1794 he was looking forward to being at Holkham from 8 January: 'I hear of great sport at Godwick.'[56] In 1783, Coke's neighbour, Sir Jacob Astley, complained that the Christmas recess from 24 December to 8 January was too short. Two days were spent each way on the journey, which allowed him little time for his 'private affairs'.[57]

While Coke's early years in politics were dominated by the American War, there were also domestic issues of concern. 'Patriots' always expressed a desire to remove corruption from public life, and a frequent target of the opposition was the number of pensions and sinecures in the gift of the king and his ministers. In 1780 a County Meeting along the lines of those proposed by Wyvill was called in Norwich to support a petition against corruption in parliament. It was attended by the two County Members, Astley and Coke, who supported the chairman, Edmund Rolfe, Coke's neighbour from Heacham, who spoke of the alarming increase in the influence of the crown in the representation of the people, the increase in expenditure, particularly on the war, and the exorbitant grants and sinecures given to the king's supporters. The Tory MP Sir John Wodehouse was amongst those who spoke against the petition. The *Norwich Mercury* reported that the petition was passed and 'we are happy to learn that much harmony and good breeding subsisted in the company who dined together afterwards at the Kings Head.'[58] Coke duly presented the resulting Petition for the Abolition of Unjust Pensions, but not surprisingly it made no headway against entrenched interests. The final years of the North government saw the increase in a movement for parliamentary reform supported enthusiastically by the opposition. While Coke spoke in parliament approving of the aims of Wyvill's County Associations, he feared that they could be unconstitutional if their delegates began to give themselves powers that only MPs should exercise.[59] Of more personal concern were his contributions to the subject of game laws and the prevention of poaching.

On 4 February 1784, Coke spoke supporting Fox and those who had formed the coalition 'from a conviction that their endeavours were laudably aimed at the furtherance of the public welfare'. He then proposed the motion 'That it is the opinion of this house that the continuance of the present ministers in power is an obstacle to a firm, efficient and united administration which could alone save the country.'[60] The vote which followed was close (223 for, 204 against) and wrecking tactics such as these succeeded in creating uncertainty through the early months of 1784. The power struggle between the Foxites and the Pittites continued until by March Pitt felt in a strong enough position to call a general election, and parliament was dissolved on 25 March 1784. Gradually public opinion, which above all wanted stability, had begun to swing in Pitt's favour. He felt he could reasonably expect the support of the reformers whose views he had frequently put forward in his parliamentary speeches. As Chancellor of the Exchequer under Shelburne he had begun work to regulate public offices and the sale of positions, which included abolishing sinecures granted for life with the Customs Department. Fox, on the other hand, by allying himself with North, a noted anti-reformer, had forfeited his radical reputation. As Pitt had calculated, both those seeking a reform of parliament and Dissenters hoping for the repeal of the Test Acts voted for Pitt and his supporters, as they felt that Pitt, rather than Fox, was more likely to achieve these ends.

Feelings against the coalition were particularly strong in the eastern counties. In Suffolk, the *Bury Post* saw the support of Pitt as 'an expedient on the way to higher things'.[61] Mr Capel Lofft, who had spoken in favour of a county petition at a meeting in March 1783 which was signed by 2,000 freeholders, asking for yearly elections and universal manhood suffrage, was a regular speaker and writer on reform. Although not himself a candidate, he advocated the support of Pitt through the newspaper's pages. By 27 March the various Norfolk candidates were offering themselves for nomination in the Norwich papers and an acrimonious fight seemed inevitable. Fox and his supporters had fallen greatly in public esteem, and Coke's close attachment to him was going to make the election a difficult one. Sir Henry Hobart of Blickling who was a supporter of Pitt and known to support the commercial interest of the city, was one of the candidates and one of the sitting Norwich members. How much time he had actually spent at Westminster during the previous session is unclear. In his letter to the electors he admitted that he had not 'attended the House of Commons because of ill health', but was 'flattered that he had been asked to stand' and begged support. The other was Coke's friend, Sir Harbord Harbord of Gunton Hall.[62] The meeting to nominate representatives for the city was held in early April and attended by more people than ever before. As well as the two sitting members defending their seats, the unwilling politician William Windham of Felbrigg stood. Although he had always spoken vehemently against the American War, he was not a natural politician and dithered as to whether to support Fox. In an attempt to distance himself from the Fox camp, he declared himself an independent who would 'never attach himself to any party'.[63] The meeting to nominate the County Members was held on 10 April in St Andrews Hall. As well as the sitting members, Sir Edward Astley and Thomas Coke, Pitt's supporter, Sir William Wodehouse, was nominated. The unpopularity of the Astley–Coke partnership was clear at the meeting. Edmund Rolfe's speech nominating Coke was frequently interrupted by cries of 'no Fox, no coalition, Pitt and the king for ever'. Astley also chose to distance himself from Fox and his old friend Coke by promising to be a 'free and independent agent of an enlightened people'. When it was Coke's turn to address the meeting 'the noise and disorder into which the assembly had now fallen prevented anything being heard distinctly.' A long board on which the words 'Astley and Coke' were painted was displayed amidst hisses and groans of the 'vast majority' and was soon broken into pieces.[64] Coke's unpopularity must have been obvious to him, and having been deserted by Astley at the last minute, he withdrew from the election. Wodehouse and Astley were duly returned unopposed, with Windham a member for Norwich. 'Thus the contest of the County was quietly terminated, which in the beginning appeared to be growing up into a strong opposition.' There were the usual parades, and toasts to Pitt, the king and the constitution were drunk.[65]

Back at Holkham, Coke licked his wounds in the knowledge that he was one of a group who had suffered as a result of Fox's miscalculations and who came to be known as 'Fox's martyrs'. In neighbouring Suffolk Sir Thomas Bunbury, who had been a County Member since 1761 was ousted and both seats went to Pitt's supporters, who had won a huge victory, taking up to 70 seats from the Foxites. Their defeat was the result of the obvious divisions within Fox's party, the unnatural alliance with his old enemy, Lord North, and the strong leadership of Pitt. Pitt's apparent commitment to reform alongside his perceived ability to provide stable government had all helped his cause. The election of Fox himself for the City of Westminster was disputed and it was announced that no result could be declared until an investigation of every vote had taken place. Rather than wait for this to be completed, Fox managed to obtain one of the seats for Orkney and took his seat on the opposition front bench on 18 May. Notions of 'patriotism' and support for reform did not prevent him taking advantage of the corrupt electoral system of the time. Meanwhile, at Holkham the letters of sympathy for Coke flowed in thick and fast, with notes from over 20 of the major Norfolk Whig families, the Keppels, the Walpoles, the Townshends and the Harbords being the most well-known amongst them.

It would be six years before Coke again entered parliament, but his first period as a County Member had been in turbulent times in which the belief that patriotism meant the upholding of liberal principals, even if this meant the loss of the American colonies, had been tested. He was not a major speech-maker, and his role in forcing George III to enter peace negotiations with the American colonies was probably not as pivotal as he claimed. However, he performed the duties of a County Member successfully by promoting the interests of his native county and its landowners. His role in parliamentary affairs was typical of that of a country gentleman of his time, but it should not be regarded as outstanding. He was one of the many who spoke little, preferred his field sports to London and had to be cajoled to come to vote. He had friends in high places, but never attained high office. It may well be that Coke's friendship was cultivated by the penniless Fox because Holkham was known for the quality of its shooting and the excellence of its library, rather than for appreciation of Coke's political abilities.

The life of an eighteenth-century MP was a very itinerant one, with many days either in uncomfortable coaches or in the saddle. His wife certainly enjoyed London society, and would frequently have travelled with him, but the opportunity to spend the sporting months at Holkham, safe from urgent letters asking him to go to London to vote, accords well with what we know of Coke.

Chapter Three

LIFE IN NORFOLK AND LONDON 1776–c. 1800

When Thomas William Coke and his wife, Jane, arrived to take possession of his inheritance at the end of April 1776, he was taking over one of the finest Palladian mansions in England. He was entering an elite group of about 400 great landed families who owned between them almost a quarter of the land of England and Wales, and whose wealth allowed them to meet the expenses of both a country estate and the London season. Their great houses were centres of political and social influence. Power was based on land, and country houses were the centres from which that power emanated.

Coke had probably not been to Holkham since his memorable visit to his great-aunt five years before, but there would have been no major changes to the house and grounds in that time. Smoke from the great open fires might well have reduced the new gleam of Kent's white and gold decoration, but the opening of the shutters and the removal of the dust sheets from the furniture under the watchful eye of the *groom de chambre* in preparation for his arrival would have revealed a well-appointed Georgian residence in perfect Classical taste.

Holkham had been built as a temple to the arts, and while Thomas had shown some interest in things Classical on his Grand Tour and brought back antiquities to add to the treasures at the hall, he knew that he could never achieve his great-uncle's understanding and knowledge of the Roman world. He vowed that he would make no alteration to a house built 'after years of thought and study in Italy'.[1] Instead he simply maintained it in good order. Clocks were serviced annually and in 1793 the pictures were cleaned at a cost of a guinea a day, amounting to a total sum of £206 – an indication of the number of paintings within the hall. The household accounts also contain payments for the regilding of picture frames. Over time, furniture needed reupholstering or replacing. In 1808 12 chairs in the steward's room needed reseating and in 1811 new looking-glasses, tables and presses were purchased. In 1804 a sum of just over £5 was spent on 'plastering and whiten-ing walls and cornices in the house' and in 1805 there was some repapering and the purchase of 'holland' (a stiff canvas) for blinds. In 1808 the north state room was repapered at a cost of £5 and Charles Martin of Elm Hill, Norwich charged £33 16s for 'modernizing furniture in various parts of the mansion'.[2] Thomas Creevey, who was a frequent visitor to Holkham, commented on the fact that little had changed there since the house was built: 'the Genoa velvet in the three living rooms has been up since the house was built eighty years ago, but fresh as four years old ... never was a house so built outside and in. The gilded roofs of all the rooms and the doors would of themselves nowadays, take a fortune to make; and his pictures are perfect, tho' not numerous.'[3]

Whether it was Thomas himself who organised these domestic affairs, or whether they were duties taken on by his wife, we shall probably never know: certainly in many households where the husband's main interests were sport, agriculture and politics, it was the lady of the house who had the cultural knowledge and design flair suited to keep the decoration up to date and in good order. For instance it was Lady Theresa Parker of Saltram rather than her husband who was writing for samples of blue damask for the saloon of their Devon home, 'as we shall soon write to Genoa and wish to fix upon the best blue for setting off the pictures'.[4]

As befitted a man whose main interests in the country were in sport and the great outdoors, it was to the park and grounds that Coke turned his attention. While the designed landscape at Holkham was not on anything like the scale of the allegorical gardens at Stowe laid out by Lord Cobham, the leader of opposition to Walpole, Holkham Park contained elements that marked it out as the work of a Whig grandee. Laid out on a formal design as suggested by William Kent and Mathew Brettingham in the 1720s and 30s, by the 1770s the setting of the house, approached through its Classical triumphal arch, built in the 1740s, with its straight avenue and complex pattern of vistas centred on the obelisk was regarded as distinctly old-fashioned. As well as the lake, a formal water feature lay to the front (south) of the house flanked by Classical 'porches'. Kent worked for many of the prominent Whig families of the time and was familiar with the ideologies to be expressed in their landscapes. The Classical style, with Italian overtones, not only linked the park to the fashionable paintings of Claude Lorrain and Poussin, of which Thomas Coke had a fine collection, but also back to the period of the Roman Republic on which radical Whigs felt English society should be modelled. They saw themselves as the 'senators of a modified Roman Republic with an hereditary German president'.[5] In the 1740s, as the building of the hall progressed and fashions in garden design were changing, a less formal element was introduced by Kent with the construction of an artificial hillock to the south of the house with a temple-like seat on it from which the house could be viewed across the lawn and basin. Clumps of trees were also planted on the north lawn. A less acceptable element of the layout at a time when exclusivity and isolation were becoming part of any design, was the kitchen garden at the head of the lake, within full view of the house. However, very few changes were made over the next 30 years, and this was the landscape through which Coke had travelled on his first visit and which was now to be developed along entirely new lines.[6]

The period between the first laying out of the surroundings of the house and Coke's return to it as its new owner had seen further changes in landscape design. Firstly, geometric layouts with vistas and straight carriageways were losing favour. Blocks of woodland planting were superseded by curvilinear clumps. Similarly, formal basins of water were being replaced by serpentine lakes, and Coke's first change was to fill in the basin in front of the house and instead concentrate on the lake to the west as the major water feature. In January 1786, William Windham was staying at Holkham and described skating on the 'pond at the back of the house ... The weather so pleasant that all the pleasure which solitary skating can give existed in great perfection.'[7] He was presumably referring to the lake before it was altered. Later in the year £219 was spent on the work of enlarging it, which included a major earth-moving exercise to create a serpentine twist at its northern end. A final fee was paid to the East Midlands garden engineer William Eames, who may have been known to Coke through his Derbyshire connections. Further work at the southern end of the lake was carried out between 1801 and 1803, by 'six men from Cheshire' whose travelling expenses as well as wages were paid in 1802. In 52 weeks, 36,000 cubic yards of earth were moved, costing a total of £743.[8]

The gardener John Sandys was employed at Holkham from at least 1781 at an annual salary of between £40 and £50, which put him on a par with senior members of the household. A Derbyshire man, he or his family may also have been known to the Cokes from Longford days. Work on the 'new gardens' involved the building of a melon pit and glazing and painting the frame, the purchase of lime and bricks and repairing the park paling. In 1782–3 £58 was paid out for new park paling and £188 for garden walls, and £433 was paid to John Sandys for the gardens and plantations account. Flower roots, bee boxes and hives were purchased.[9] The following year a further £481 was paid to Sandys, with £60 for 'garden buildings' in 1784 and £511 in 1786. His major contribution, however, was the creation of woodland. In 1781 he planted over 7,000 trees in 22 acres near the East Lodge. The majority of these were hardwoods such as oak, ash and beech, but fast-growing conifers were also included as nursery crops. Ten acres near the old kitchen garden by the lake was also planted up with 3,000 trees, as well as four acres on the marsh. A further 40 acres and more than 11,000 trees were planted in 1784. Between 1785 and 1789 Sandys was responsible for planting a further 179 acres within the park with 396,750 trees.[10] He continued to supervise the planting of the new woodlands and the running of the gardens until 1805, when he married a widow from Wells. In 1806 a new gardener, James Loose, first appears in the accounts. However, Sandys continued to be an adviser in forestry matters.

A general development during the second half of the eighteenth century was the increasing size of parks, thus isolating the owner from the local population and emphasising his increasingly powerful hold over society. As the gulf between rich and poor increased, so did the perceived importance of seclusion. Although it is

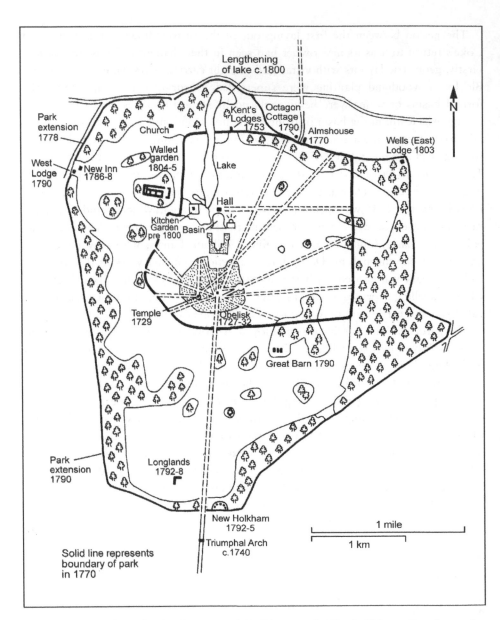

Map 1: Plan to show the development of Holkham park. The bold inner line shows the extent of the park when Thomas William Coke inherited. Most of the vistas shown disappeared during the following 60 years. Most of the clumps and belts of trees surrounding the park were planted during Coke's time as owner.

probable that the old village at Holkham had all but disappeared before Thomas Coke had begun building on a site adjacent to the old manor house, several farms remained within the vicinity. The great avenue from the triumphal arch to the obelisk was bordered by fields, worked by the tenants of Longlands and Honclecrondale farms, whose houses and buildings could be seen from the carriageway. As the leases of farms beside the avenue fell in, their lands were incorporated into the park and fields worked as part of the Home Farm which, by 1816, covered at least 2,000 acres within the park. While the houses of tenants were not to be tolerated within sight of the hall, arable fields were acceptable. The enlarged park was enclosed, firstly by iron palings, and between 1830 and 1839 by an 8-mile brick wall and a belt of trees, but remained in agricultural use. To ensure further privacy, the coast road had to be rerouted, so that by 1800 the park covered 3,500 acres, thus bearing comparison with the creations of some of the most powerful aristocrats of the land: an indication of the way visible divisions between the old nobility and the more wealthy of the gentry were becoming blurred. Once the park was enclosed more tree planting could take place, as much as anything to provide cover for pheasants, so that Coke could indulge his passion for field sports on his own doorstep.

The first reference to fees for the country-house architect Samuel Wyatt appears in the household accounts in 1780, and between 1799 and 1805 he was responsible for the building of the new lodges which were necessary at the entrances to the extended park. In July 1796 the landscape artist and man-about-town Joseph Farington met Wyatt, who 'has been much employed by Mr Coke of Holkham and spoke highly of him'.[11] Wyatt also designed the kitchen garden, laid out between 1800 and 1806 to replace the old one at the head of the lake, now considered far too near the house. The new site to the west of the earlier one was completely hidden from view and covered six acres. Well-supplied with forcing houses, it was admired by early nineteenth-century visitors, who thought it the finest they had seen. Its focal point was the vinery, which was described as a 'graceful little paradise', to which Coke liked to take his visitors on a summer evening and where, surrounded by the vines, exotic flowers and fruit trees, they could sit and drink tea.

A curious entry in the accounts for 1802 records the building of a 'subterraneous passage to the garden'. It is most unlikely that such a passage could have run all the way to the kitchen garden, although it is known that such passages did exist, enabling servants to stay hidden from view when fetching fruit or vegetables, or going to hang out laundry.

Coke's new work in the park included not just the decorative lodges, vinery and kitchen garden, but also practical farm buildings. Particularly impressive, and again designed by Samuel Wyatt, is the Great Barn, which bears comparison with other estate farm buildings designed by him in Essex and Cheshire.[12] Built in the 1790s, it was one of the buildings visited during the sheep shearings. Standing on rising ground just off the great avenue, it could have been clearly seen from the hall; equally those visiting the barn would have glimpsed the hall beyond. The link between landlord and agriculture was obvious, and the dual role of the patriot as patron of art and agriculture was thus visibly demonstrated.

By the time Coke began work at Holkham 'Capability' Brown was dead, but many of his followers were very active. Coke chose to employ men who were local either, as in the case of Wyatt, Sandys and Eames to his Derbyshire estates, or, as in the case of Humphrey Repton, to his Norfolk one. Born in Bury St Edmunds and firstly moving to Sustead near Aylsham he became a close friend of William Windham, spending much time in his library at Felbrigg, where he was introduced to some of Windham's learned friends such as the botanist and President of the Royal Society, Sir Joseph Banks. Windham returned the visits: 'he says he likes my snug study at Sustead better than the old rambling library at Felbrigg.'[13] In 1783 he accompanied Windham during his brief time as Lord Lieutenant of Ireland, returning to run his election campaign in 1784. Repton was to go on to become one of the more famous of English landscape gardeners (a term which he himself invented). However, his first commissions were Norfolk ones, working at Catton, near Norwich, for Jeremiah Ives, and at Holkham.

Coke was still not entirely satisfied with the setting of the lake and in 1789 Repton was commissioned to suggest improvements. Like his immediate predecessors, Repton aimed to create informal and 'natural' landscapes, and his main proposal was to establish an elaborate walk providing a variety of views along the eastern shore of the lake. These included a boathouse and fishing pavilion as well as a chain ferry leading to a 'snug thatched cottage' on the far side. This west side was to contrast with that on the east, which was to be more 'rural' but not 'neglected'. 'Nature may carelessly stray in her rustic garb, while Art, unperceived, goes before her at a distance and clears away some thorns and briars that would obstruct her narrow path.'[14] A unique aspect of Repton's work was the production of a beautifully bound 'red book' for his clients. 'I wanted the means of making my ideas equally intelligible to others. This led to my delivering reports in writing accompanied by maps and such sketches that at once showed the present and proposed portraits of the various scenes capable of improvement … the effect produced by my invention of slides made the sketches interesting.' The 'before' and 'after' views, which were overlaid, with the upper version sliding away to show the lower, were a highly successful marketing idea, although Repton did realise that many of his drawings were 'only required to shew to friends without (the clients) ever intending to do anything to their place'.[15] This seems to have been the case with Holkham. The red book in which Repton's scheme was presented is elegantly written and beautifully illustrated, but probably very little was done. In 1802 money was spent on walks which may have been suggested by his plan, and it is possible that the 'subterraneous passage' was part of this scheme, but there are no references in the accounts to a chain ferry, the cottage or boathouse. The lake was not only an ornamental feature. It was stocked with fish, and in 1798 a 'new pleasure boat for the lake' was built. The lake could also be used for winter sports when it was frozen. In 1799 Coke's son-in-law, Lord Andover, had reason to be grateful to his servant, who 'had saved his life from an accident' while skating on the lake.[16]

Coke's interest in transforming the surroundings of the hall into an informal park is clear. As a sportsman, he wanted to extend the planting of clumps as pheasant cover, but also, as an agriculturalist, to maintain farming within the park boundaries. His anonymous biographer wrote of the farms amalgamated with the park in the 1770s: 'These occupations now constitute a great portion of the highly cultivated parts of Holkham Park.'[17] While it was not unusual for some farmed land to remain within parks, the extent and its proximity to the main entrance along the avenue at Holkham was commented upon by visitors. Noting the mixture of lawn and cultivation, the Cumbrian agricultural improver John Curwen wrote: 'What can be more beautiful than the diversified scenery that there presents itself ... the effects of order and industry, combined with abundance, must be gratifying to every spectator.'[18]

Most of the work on the park was completed by 1810. Trees continued to be planted at a rate of 50 acres a year until a shelter belt had been created all around the park wall, and Coke and the family could simply wait for his work of planting to come to maturity and enjoy the landscape they had created.

The all-important consideration as the park was extended and planned was the preservation and encouragement of game. By the end of the seventeenth century the provisions of the game laws had ensured that the shooting of game was confined to a very limited social class. An Act of 1671 laid down that game could only be taken by those owning freehold property worth £100 a year or long lease and copyhold property of an annual value of £150. In an effort to prevent poaching an Act of 1755 prohibited the buying and selling of game. Game was thus regularly given as gifts, and Farington recorded in October 1807 that the departure of the London coach from Fakenham was delayed while waiting for game from Holkham.[19] The real increase in the popularity of the sport came in the second half of the eighteenth century with the improvement of guns, as the cumbersome matchlock was replaced by the safer and more manageable flintlock.[20] Further improvements ensured that ever-larger numbers of birds could be shot, and as guns became lighter birds were taken on the wing, rather than on the ground. The ability to hit a flying target was the mark of a good sportsman. While partridges were birds of the open country, pheasants preferred a more wooded environment. Young woodland was thought to be best, especially if it contained larch trees, 'on account of its branches growing nearly at right angles to the stem: this renders the sitting position of the birds very easy'.[21] However, when the trees reached maturity, it was necessary to introduce a shrub layer of some shade-tolerant species such as bird cherry. Pheasants also prefer the edges of woodland to the deep centres and so belts, sinuous plantations, and small clumps were the most suited to game cover. It is precisely this sort of arrangement which was typical of gentlemen's parks, such as Holkham, in the late eighteenth century.[22]

Coke's enthusiasm for field sports from childhood has already been noted. When his father became the owner of Holkham it was the sporting possibilities that attracted him to the area, and when he inherited he was loath to give up the pleasures of the field for a parliamentary career. However, as we have seen, in spite of the fact that parliament frequently sat throughout the shooting season, Coke did manage to combine visits to London with periods at Holkham.

While guests were regularly entertained at Holkham, Coke also shot on other parts of the estate. Godwick was always a popular venue, and in 1784 Thomas Whisker was paid for attending Mr Coke there. He would stay at Godwick Hall, as a guest of his friend and protégé, the Revd Dixon Hoste, rector of Tittleshall. Isaac Serjant was paid £10 a year in the 1780s to take care of the woods and game. Coke also shot at Castle Acre, where rooms at the Ostrich inn were described in an inventory as being kept aside for him.[23] He also took guests to this more distant part of the estate. Joseph Farington wrote in his diary for 16 September 1795: 'Charles Fox and Mr Coke this day shooting at Castle Acre.'[24] Pheasants were being encouraged on estate land at Fulmodestone Severals, where buckwheat was provided for them. Not only did Coke shoot and entertain at Holkham, but he also spent days away with other members of the local gentry. When Thomas Moore called to see his landlord in October 1805 about his lease, he was asked to come again after dinner, Mr Coke 'being gone to Houghton a-shooting'.[25]

However, it was around Holkham that most of Coke's sport was found, and the number of partridges, hares, rabbits, pheasant, woodcock and snipe shot, as well as who shot them, were noted in the game books. Samuel Wyatt was responsible for the creation of a game larder lined with alabaster and provided with iron racks and hooks for hanging the birds. Partridges were by far the most numerous birds, with between 1,300 and 2,500 killed most years. Numbers of pheasants varied from 260 to 858, with the much larger number of 1,443 shot in 1834. The numbers were gradually increasing over the years as the improved guns allowed for faster shooting and birds were preserved and fed by gamekeepers to be driven by beaters towards the guns. In 1822 Coke's youngest daughter, Elizabeth, wrote to her future husband of the huge number of birds shot – 800 in one day: 'George Anson killed 130 with his own gun.'[26] Shooting started in late October and continued twice a week during the season. Interest in extending the variety of birds on the estate is shown by the fact that French partridge eggs and golden pheasants were bought in 1802. In the 1797 season there was shooting every other day, with three or four guests, one of whom was Horatio Nelson, presumably staying with his father, the rector of nearby Burnham Thorpe. In November 1797 the Prince of Wales was shooting at Holkham and Houghton.[27] More frequent guests were leading Whig politicians such as Charles James Fox and Richard Sheridan, and aristocrats such as the Duke of Bedford and Lord Spencer, as well as relations, such as his brother-in-law Lord Sherborne and, later, his daughters' husbands, Lord Andover and Lord Anson. Local dignitaries such as William Windham and members of the Colhoun, Rolfe, Wodehouse and Walpole families were also regular visitors. All these, except Sir John Wodehouse, were Whig sympathisers, and we will never know how often conversation turned to the parliamentary battles of the day and how many parliamentary tactics were mapped out in the woods around Holkham or the great rooms within the hall. Writing from Holkham in October 1794 as the Whig party was tearing itself apart over its attitude to the French Revolution, Fox stated that while men such as Coke, the Dukes of Bedford, Guilford and Derby, 'and some others with myself, make undoubtedly a small basis, but then how glorious it would be from such small beginnings to grow into a real strong party, such as we once were'.[28]

Notably absent from the lists of shooting guests are members of the tenantry, with General Fitzroy of Kempston the only exception. Like all landlords, Coke retained the shooting rights across his farms, and Thomas Moore, tenant of Hall Farm, Warham, recorded in October 1807 how 'Mr Coke and a party of lords and gentlemen came to shoot.'[29] The farmers were frequently given gifts of game, but the sport itself was strictly limited to Coke's landowning friends. There is no record in the game books of a tenants' shoot at the end of the season, which was a feature of other estates at the time. Moore described how 1,457 head were shot in the three days preceding 26 October 1807 at Holkham. Coke had shot 60 on the first day, 66 on the second and 70 on the third, but whether Moore was there or reporting hearsay is not clear.[30] Significant incidents are recorded in the game books. In 1813 Mr Lambert killed a pheasant and a hare at one shot, while in the same year Mr Acland killed a 'walking cock like an ostrich and a hare like an elephant'. On 7 September 1820 two bustards were shot – surely some of the last of a species which survived longest in Norfolk, but was declared extinct in the United Kingdom in 1838. Most memorable was the killing of two woodcocks with one shot by the famous sculptor Francis Chantrey in November 1830. The two woodcocks were then delicately carved on a marble tablet by Chantrey and presented to Coke with an inscription recording the event. The tablet now hangs in the corridor above the great hall.

The only fatal shooting accident of which there is any record is that involving Coke's son-in-law Lord Andover in 1800. Setting out late with his servant to follow the shooting party he stopped on South Creake Common, cocked his double-barrelled gun ready to shoot and handed it to his servant while he bent down to catch his dog. One account suggests that the servant was on horseback and the horse suddenly swerved. By some accident, the gun then went off and Lord Andover was shot in the back. He was taken to a nearby farmhouse, where he died, insisting that no blame should be attached to the servant.[31]

This obsession with field sports was not to everyone's taste. Coke's youngest daughter, Elizabeth, wrote to her future husband, John Stanhope, in 1822, 'Your conversation will be more than usually delightful to me. I am so tired of mangled hares and missed woodcocks.'[32] Coke himself appears to have continued shooting until the last year of his life.[33] In a letter of 1839 Edward Ellice, a frequent visitor to Holkham in later years, wrote to Fox's nephew, Lord Holland, that Coke was 'out with the shooters from ten to sunset every day'.[34]

Alongside shooting went a passion for fox-hunting. Not only did Coke keep hounds at Holkham and on his small mid-Norfolk estate at Elmham, but he also had kennels at Epping to allow for a day's hunting when he was in London. In 1794 he paid a surgeon to attend his huntsman in Essex when he broke his leg. Stables of hunters were kept at both Elmham and nearby Tittleshall. Expenses for hounds first appear in 1785, when he moved a pack from his brother-in-law's home at Sherborne, Gloucestershire, to Holkham. In 1793 fox-hunting expenses amounted to £275, while in 1794 they were £506, including £60 for earth-stopping in Norfolk and Essex. In November 1789 the scholarly Windham was fox-hunting at Holkham. He set out from Felbrigg on horseback, thinking through a mathematical problem encountered in an academic paper he had been reading. He stopped in Holt, and while his horses were being baited 'had the satisfaction of hitting off a mode of reasoning, which the result showed to be right. ... I should not have had so much pleasure in the fox-chase this morning, if I had not hit upon this solution of the case contained in that paper.'[35]

This preoccupation with field sports was usual amongst the landed gentry, and a fierce application of the anti-poaching laws, strengthened in 1770 and 1773, was typical of the period. We know of very few speeches that Coke made in the House of Commons, but as we have seen several of these were in connection with the game laws. This attitude could make him enemies, and may have been one of the reasons why he lost the 1784 election. 'There is no doubt that his overbearing ways and fondness for game preservation did him more harm at the polls that his services to agriculture did him good.'[36] There are several records of Coke paying constables for taking poachers to Walsingham Bridewell. Although it is unlikely that the electors would have had much sympathy for the poachers and would have approved the harsh treatment meeted out to them, field sports could make him unpopular with tenants near his favourite haunts, where the protected game attacked their crops.

But the fact that he spent all day in the saddle or out with a gun did not mean that he neglected his other duties. John Farington, who was a close friend of the portrait painter Sir Thomas Lawrence, recorded that Coke had told Lawrence, when sitting for his portrait, that 'he does all his business in the morning between seven and ten after which time he is at leisure to shoot or for any other amusement. He says that he makes it a rule to answer every letter the day he receives it.'[37] Coke's claim to have conducted all his affairs himself is somewhat exaggerated. Samuel Brougham kept the household accounts in the 1790s, while Francis Crick took over in 1801, and it is clear that much of the day-to-day contact with tenants from that date was through Crick. However, when the salaries of these men (£120 and £300 a year) are compared to the much higher one paid to Francis Blaikie, who took over as agent in 1816 (£550 a year, rising to £650), it seems likely that their responsibilities were more limited than his.[38]

Great houses were designed for pleasure as well as power, and while most visitors to Holkham in the early days of Coke's ownership came for the shooting, the house itself was the centre of hospitality. A true patriot could be recognised by his open house and entertaining. This suited Coke's extrovert personality, as well as being crucial to maintaining his political popularity. As we will see, this reached a climax in the celebrations of 1788. Guests plus their servants would be entertained for the shooting and for the sheep shearings, and in several years he found it necessary to hire beds from Wells to accommodate all his visitors. The fact that the house was designed primarily for entertainment is indicated by its design. The central core contained the public rooms such as the grand saloon and two drawing-rooms on the south side, state bedrooms to the east, the statue gallery to the west and the dining-room to the north, while the four pavilions each had the separate functions of chapel, kitchen, strangers' and family wing. Coke's private quarters were in the family wing, which had its own dining-room as well as rooms for the footmen, valet, secretary and maid at ground level, with the library and living- and sleeping-quarters above.

As well as containing a simple Classical family chapel panelled with marble below and decorated mainly with Italian Renaissance paintings, the chapel wing also housed the nurseries, where the couple's children would probably have lived with their nursemaids.

Guests and their servants were lodged in the strangers' wing. Visitors had their own private quarters, usually a bedroom and dressing-room which they also used as a sitting-room, only mingling with the family and other guests in the more formal state rooms at the centre of the house. The studious William Windham remarked that 'A cell in a convent is not a place of greater retirement than a remote apartment in a house such as Holkham.'[39] In many ways it was a reasonably comfortable house in which to stay. The more intimate arrangements in the wings in contrast to the grandeur of the state rooms allowed for some informality. Arthur Young expressed his enthusiasm after his visit in 1768, eight years before Thomas inherited: 'so convenient a house does not exist – so admirably adapted to the *English* way of living and so ready to be applied to the grand or comfortable side of life.'[40] As well as the interconnecting bedrooms and dressing-rooms with both central and back staircases allowing access for servants in the family and strangers' wings, there was also the innovative luxury of water closets, two of which were in the strangers' wing.

In this Holkham and its designer, Coke's great-uncle, were ahead of their time: '2 Marble Stools for water Closets' costing £9 5s each and a further one for £7 10s were installed in the family wing in 1741. Two in the strangers' wing another off one of the state bedrooms and a double-seater off the hall apse were all installed later.[41] In 1803 £116 was spent on the 'engine house, new larder and water closets'. When Thomas Creevey visited Holkham over Christmas and New Year 1837/8 he caught a cold and so was confined to his room for several days. He stayed in a 'charming bedroom on the ground floor with a door at hand to go out of the house if I like it and another equally near for nameless purposes'. He was impressed by the way he was looked after. A maid came and lit his fire at seven and then kept it stoked up almost hourly and 'my water is in my room at eight'.[42]

Although the library was situated in the family wing it was generally open for the use of visitors. When Windham visited Holkham in August 1792 to meet Sir Joseph Banks he spent some time in the library reading Lord Bolingbroke's *Letters on the Spirit of Patriotism* and *Dionysius Italicarnassus*. 'I had left a paper in this book from the time of my being at Holkham two years before, I believe.'[43] The American Richard Rush described his time spent in the library while attending the sheep shearings.

> The library of many thousand volumes is a treasure and ... on one of the days that I was entertained, during the short interval between the morning excursions and the dinner hour did I catch stragglers of house guests, country gentlemen too, who had not been out to the fields and farms at all, though they had come all the way to Holkham to attend the sheepshearing! And no wonder. In fact they were of the younger portion of the guests, not long from the university, so recently that the love of practicality inspecting wheat fields ... had not so deadened classical ardour as to keep them from stealing off to where they could find curious editions of Pliny and Ovid and the Georgics, or if they liked Italian better, lay their hands on the identical Boccacio which Cosmo de Medici sent as a present to Alphonso, King of Naples.[44]

However, George III's younger son, the Duke of Sussex, was not so lucky. His first visit to the sheep shearings was in 1821, but as his interests were more literary than agricultural, he made for the library, only to find that it had been 'converted into bedrooms' to house some of the many guests.[45]

There is no doubt that Fox too would have enjoyed the library. His correspondence, particularly with the Classical scholar and tutor Gilbert Wakefield, who spent several years in Dorchester gaol for writing what was regarded as a 'seditious pamphlet' in 1798, showed his knowledge and interest in Classical literature. There was certainly much to interest him in Coke's fine collection.

We know that Coke was no bibliophile. His knowledge was described as 'practical' rather than 'book', and his love of sport at school had 'well nigh obliterated his love of study'[46] When the poet and antiquarian William Roscoe visited Holkham in 1814, he found that many rare manuscripts and printed folios had simply been dumped in an upper room 'in consequence of their unsightly condition'.[47] However, Coke did encourage scholars to use a catalogue of his books, and he bought topographical, religious and agricultural books as well as popular novels, which joined the Classical works on the shelves. His admiration for America and George Washington had led him to buy two books by him, *Letters to Sir John Sinclair on Agriculture and other interesting topics*, and *George Washington to the people of the United States announcing his Intention of retiring from Public Life*, both published in 1800. About 400 pamphlets, which he either bought or was given, range across a wide variety of subjects, including agriculture, Catholic emancipation and the ending of slavery. He subscribed to the local newspapers, and copies of Arthur Young's *Annals* were purchased. A regular subscription was also paid for the *Farmers' Journal*.

The smooth running of the household relied on an army of servants, and for most of Coke's life there could have been as many as 40 of them living in the house, but hidden from view. The house was so arranged that backstairs and corridors allowed them to go about their work unseen, and it was this aspect of the house that particularly impressed visitors.[48] It was typical of house design at the time for servants to occupy the lower or 'rustic' floor, which was almost exclusively the servants' domain with its own hierarchy. The arrangement of rooms allowed for the upper servants' offices to be placed under the principal apartments and so conveniently near 'master and his company'.[49] As close as possible to the family wing on the south side were the housekeeper's and butler's rooms. The butler's suite consisted of pantry, plate-room and bedroom. Immediately across the corridor were the stairs leading to his cellars. The housekeeper was at the other end of the south corridor, and from there she supervised the still–room, where cordials might be made and preserves and cakes kept. The laundry and the dairy were below, at the cellar level. Nearby on the rustic floor was the dessert-room where sweetmeats, biscuits and other delicacies were made and stored. This was the domain of the confectioner. The servants' breakfast-room was on the east side, alongside the steward's rooms, which connected with his store and the servants' dining room on the north front. The servants' hall, where up to 25 servants might eat, was next to the kitchen. Here meals would undoubtedly have been hotter than those taken by the family and guests in the distant dining room. Stairs next to the servants' hall led up to rooms at the top of the kitchen wing which were for the use of the male servants and footmen, while the maids slept in the lower floors. The kitchen was huge and impressive, with its ranges and glinting copper pans. A new 'cast-iron steam kitchen' was installed in 1804, but needed repairing in 1813. Here the cook was in charge and in the 1830s a Frenchman, Philippe La Roche, was employed. Intriguingly, at this time Coke bought a copy of a poem entitled 'Waterloo' from 'Mrs Roche of Dover'. Was she related to the cook?[50]

Whilst most of the servants went about their work unseen, some, such as the porter, butler, footmen, postilions and coachmen, were in public view and so were provided with liveries. In 1784 £4 4s was spent on trimmings for liveried servants. In 1785 eight livery hats were purchased, and in 1804, 'Cheadle hats for liveried servants'. Again, in 1805 £135 was spent on liveries. The house porter had a room next to the west front door. From this door, guests could be shown directly to the strangers' wing and were often received here rather than at the central door leading to the marble hall. In common with many great houses, the posts of house steward and butler were sometimes combined, and he, along with the cook, would be the most highly paid servant, commanding in 1806 a salary of £100 a year. The master of horse was paid £55, while both the *groom de chambre* and Coke's valet were paid £42. Most other servants earned between £10 and £25, with the housemaids, of whom there were six, earning eight guineas, sums which are roughly twice those paid by Coke's great-uncle in the 1720s.[51] The total cost of servants' wages in 1786 was £950. The amount paid by Coke in luxury tax gives some idea of the size of his establishment. In 1808 he was taxed for 26 male servants, 21 riding horses, three four-wheeled carriages and 28 dogs.

There is little information about the management of Coke'e stables. In 1786 both horse stables and hunters' stables are mentioned in the accounts[52] and coachmen, postilion, huntsman and whipper-in were all employed on a regular basis. We know something of the number of horses and carriages kept from the amount paid in luxury tax. If the figure given is to be believed, the number changed very little over Coke's lifetime, with for instance 21 carriage horses and 61 husbandry horses recorded in 1803. In 1838 18 horses for 'riding and drawing [driving?] and 4 ponies' are listed.[53] Horses were not kept in London because of the expense, but were returned to Holkham once they had pulled the coaches containing the family and wagons of goods to the city. The coach stayed in London and horses were hired to pull it when needed.

Until her death in 1800, the management of the house, and with it social life, was the province of Coke's wife, Jane, who, by all accounts, was a good hostess who enjoyed both London society and entertaining at Holkham. By the 1780s the majority of servants in great houses were female, and the lady of the house was responsible for them. Although the *groom de chambre* and housekeeper would be in charge of the day-to-day running of the establishment, with up to 40 servants employed at any one time, Jane's role was very much that of a manager today. The amount of personal interest in domestic affairs taken by the lady of the house could vary considerably. Jane's predecessor, Lady Margaret's, attention to detail was described by a visitor. 'She never misses going round this [the kitchen] wing every day', and she was spied by one guest in the kitchen at six in the morning. It was Lady Leicester's personal care that had resulted in the excellent dinners given at Holkham.[54] The fact that this was commented upon suggests that such a level of involvement was unusual, but there is no doubt that a well-run household relied heavily on the ability of the woman in charge. When Coke's youngest daughter was to be married, she sought advice from her older sister, Lady Andover, on the sort of person to look for to fill the post of housekeeper. The not-altogether-serious reply described a paragon:

> She must not have a will of her own in *anything*, and be always good humoured and approve of everything her mistress likes. She *must not* have a good appetite or be the least of a *gourmand* or *care* when or how she dines, how often disturbed, or even if she has no dinner at all. She had better not drink anything but water.
>
> She must *run quick* the instant she is *called*, whatever she is about. Morning noon and night she must not mind going without *sleep* if her mistress requires her attendance. She must not require high wages nor expect any profit from the *old clothes*, but be ready to *turn* and *clean* the *dirty gowns*, not for herself but for her mistress, and then sell them for an old song as she is *bid* and be satisfied with two gowns for herself. She *must* be a *first-rate* vermin catcher.
>
> She must be *clean* and *sweet* and *very* quick. She must have ears (strong ones), eyes and hands, but as for thinking or judging for herself or being in any way independent (if especially her mistress be a Whig or of liberal *principles*) she must not think of such a thing; and let her not venture to make a complaint or difficulty of any kind. If so she had better go at once.
>
> She may gather up gossip as much as she likes but never *tell any*.
>
> Implicit obedience, the first essential; extra-ordinary disinterestedness, united with a love of strict economy, the second. Honesty that will bear the closest inspection; unceasing activity; unimpeachable good health and extreme good humour *indispensable requisites*.
>
> She must in short, do everything, gain nothing except the few pounds she gets for her wages and be alive to the fact that she has a very good place.[55]

Elizabeth Coke could probably not hope to find such a person, and her more realistic requirements for a housekeeper were far simpler. She wrote to her future husband, John Stanhope, 'Pray do not engage a terrible looking housekeeper, if she frightens you she will frighten me much more. I want a goody sort of person who will occasionally make up a broth or sago for the poor people.'[56]

How far Frances Trinder, the housekeeper of Coke's sister Elizabeth Dutton, came up to the standards required by Coke's daughters is not clear, but when she died in 1844 aged 75 after 40 years' loyal service, Elizabeth's son and daughter-in-law put up a handsome tombstone to her in the churchyard at Sherborne. The inscription commemorates her 'upright and Christian character' and her 'faithful and affectionate service to the end of her life'.

Since the hiring of senior household servants would probably have fallen to Jane Coke, most of them would have been entirely unidentifiable to Coke. However, his steward and farm bailiff were men with whom he would have conversed and for whom he would have had respect. When Edward Wright died in 1809 after 17 years as farm bailiff, Coke put up a marble monument to him in Holkham church at a cost of £13 8s 10d. Wright was a popular figure amongst his fellow farmers, and Thomas Moore of neighbouring Warham was a pallbearer at his funeral.[57]

The picture we have of Coke is of a country gentleman living in the style of an aristocrat. His Palladian mansion and his retinue of liveried servants gave him all the trappings of his titled neighbours, Lord Walpole and Marquess Townshend. The size of his estate was more than twice that of any other in Norfolk, and his way of life and the scale of his hospitality eclipsed that of other local landowners. Stories of offers of ennoblement which he refused are almost certainly apocryphal, although Farington claimed to have heard from the Revd Mr Hamond, who had been told by the Revd Mr Dixon Hoste, who had seen the letters, that Fox had offered him a peerage.[58] No doubt he enjoyed the nickname 'King Tom', and being described by the Tory, George Canning, as 'a landed grandee'.[59] However, whatever stand he took 'on the popular side of politics with Charles Fox', he was, according to one of Joseph Farington's informants, 'himself thoroughly an aristocrat wishing to dominate over the minds of others and to be looked on as a ruling chief'.[60]

To earn the character of 'patriot' only a limited amount of paternalistic charity was required, and Coke's charitable gifts were small, although he gave to a wide range of recipients. The days had long gone when a great landlord was expected to provide largesse for the poor, and Mrs Stirling records that he ended the distribution of food from the hall to beggars which had been customary in the dowager's day.[61] Sums of not more than £5 were given to 'the poor of Holkham and the poor at Wighton workhouse', but other than this, his gifts were confined to individuals who may well have been known to him as ex-employees. Widow Custance received an 'annual charity' of five guineas and Widow Shepherd, 'late of the Feather's Inn, Holt', one guinea. In 1796 he gave Christmas boxes to the Wells ringers, the post boy, a blind man in Walsingham and the book-keeper to the London coach at Fakenham. The wife and children of Benjamin Able received 10 guineas via the Weasenham tenant Thomas Sanctuary. In 1804 £1 was given to John Hancock, 'who has lost a cow', and Thomas Breame, who was 'turned out of his farm'. In 1808 'Margeson of Wells', who lost his house in a high tide, was given 10s 6d and Widow Woods, who had lost her husband at sea, was given £5. A tailor in Fakenham was given a guinea charity. Coke also gave money to sailors from Wells. In 1800 sailors 'who had suffered shipwreck' were given five guineas, and 'a sailor in Wells injured by a boat falling on him' received a guinea. Medical expenses were also sometimes met. The apothecary's bill for Mrs Mary Lee of Hingham amounted to the large sum of £24 5s 2d. Although apothecaries were at the bottom of the medical hierarchy, below surgeons and physicians, the shortage of trained physicians meant that they frequently performed more than their formal role of dispensing drugs and potions. The travelling expenses to London for John Harman to seek advice over his bad eyes were paid in 1804. The nursing of John Sewell 'until his death' was paid for, as was that for Robert Dodsworth, the Holkham baker. Funeral expenses were also sometimes covered.[62]

It seems that Coke was occasionally appointed as a trustee for small sums of money. The Holkham Friendly Society had entrusted him with £200 and every year he paid out interest of £10 to them. In 1799 he paid Ann Buck £2 12s 6d, being 'part of the principal money left by her husband in the hands of Thomas William Coke'. A similar sum was paid out over the following years. The father of William and John Cox had left £200 for their upbringing, and this was used through the 1790s for their clothes and education. William Cox was instructed as a cook and worked in the Holkham household for several years. When he left in 1802, he was given £10. Coke was also giving an annuity of £21 to the widow of a former tenant, possibly also under the terms his will. Occasionally he contributed towards the cost of education. In 1784 he gave five guineas to the Revd Mr Molden and in 1803 five guineas to Widow Williams for the education of their sons.

As well as outdoor activities there were the usual country and indoor amusements to fill the family's time when they were at Holkham. Guests found that it was possible to combine scholarly activities and a quiet retreat to the library with companionship. Windham described the pleasure of staying at Holkham in a dairy entry for January 1786:

> During the whole of my stay I enjoyed myself very much; in this enjoyment the house itself has no small share. Of the mode of existence that varies from day to day, none is to me more pleasing than habitation in a large house. Besides the pleasure that it affords in the contemplation of elegance and magnificence, the objects it presents, and the images it gives birth to, there is no other situation in which the enjoyment of company is united with such complete retirement.[63]

Days at Holkham would follow a standard pattern, whether or not there were guests. Typically, breakfast would be at nine-thirty, with dinner at four. Between those meals the gentlemen went hunting, shooting or perhaps for a farming ride either around the Home Farm or to a nearby tenant. When Windham visited Holkham in August 1792 he and Coke rode in the morning, 'Coke lending me a young horse of his'. Later, in 1809, he and Coke went for a 'long ride' in the morning. They inspected some fly-bitten sheep and the threshing machine, as well as riding through the woods, where Coke promised the enthusiastic tree-planter some cuttings of sea buckthorn, French willows and evergreen oak. Windham noted in his diary that he had forgotten to ask the principal properties of the oak.[64] The ladies were more likely to go for a walk or ride in the park, or to read, draw, play or embroider in one of the smaller drawing- or dressing-rooms, or the library. Dinner would be a formal affair, after which the ladies would withdraw, to be joined by the gentlemen later for cards, reading or possibly dancing in the drawing-room. At a dinner at Houghton in October 1807 attended by Coke and many of the leading Whigs, including Lord Spencer and the Duke of Clarence, the duke 'kept the company at table until nine – a thing very unusual at Houghton as Lord Cholmondesley generally goes to the ladies in half an hour after they retire'.[65] Supper would be a light meal taken about ten.

The making and receiving of visits from neighbours was now much easier, as both the roads and the design of wheeled transport improved. Sir Jacob Astley of Melton Constable told Farington in 1806 that he never slept at Holkham when he dined there, but always made the 10-mile journey home, something that would have been far more arduous 50 years earlier.[66] Coke would also go visiting, and Windham noted in his diary that 3 July 1792 was the first time that they had dined together at Felbrigg.[67] For those from further afield, the journey from London could now be made by stage-coach, either to Fakenham or Norwich, in about 17 hours.

The local towns of Wells and Walsingham also both provided entertainment. 'No song, no supper' was performed in Walsingham Theatre in January 1799 and much enjoyed by several Holkham tenants.[68] Mr and Mrs Coke were patrons of the theatre in both towns. 'Bespoke' nights were typical of the period, when for a sum of about £10 members of the local gentry would request a particular play. Their name would appear on the posters and they would attend, often accompanied by a large party from the mansion. In July 1779 *The Clandestine Marriage and the Padlock*, written by Colman the Elder and Garrick in about 1766, was performed at Walsingham at the request of the Cokes by a company known as the Leicester Company.[69] Cock-fighting was a sport enjoyed in Wells, but probably by the tenant farmers rather than their landlord. In February 1805 Thomas Moore went to a fight followed by a dinner in the Fleece for 52 people.[70] On wet days guests at Holkham might be amused by cock-fighting under the portico – a sport much enjoyed by the Duke of Sussex.[71]

Kings Lynn was the nearest large social centre to Holkham, and the Cokes attended events such as the annual Mayor's Banquet. In September 1784 Arthur Young and his two young aristocratic charges, the La Rochefoucauld brothers, on an agricultural tour from France, were there and they described how much was drunk, 'this is the custom here, and what is more it goes on for a very long time'. Young introduced them to 'Mr Coke, the owner of the finest house in England which we are anxious to see'.[72] In December 1807 William Windham and his wife visited Holkham from their home at Felbrigg. From there they went to a ball in Lynn with Coke's daughter, Anne Anson, as well as 'Mr Coke's neice Miss Blackwell, with another lady of the Holkham party, viz. Lady Ponsonby'. Coke himself did not go, having fallen when attempting to jump a ditch.[73]

Coke also spent several days at a time in Norwich. The journey took three or four hours and there would be friends to visit, theatres to attend and business to conduct, especially at electioneering time. The Gregorian Club, with its Masonic-style ceremonies and insignia, was frequented by the Whigs. Its chief patron was Sir Edward Astley, father of Sir Jacob, and Coke was its president in 1778 and invited its members to visit Holkham in July 1799. Popular with both town and country gentlemen, the club was often a venue for heavy drinking rather than serious politics. It closed in the early nineteenth century.[74] A sport much enjoyed by the gentry on visits to Norwich was bull-baiting, and both Coke and Lord Albemarle were patrons of a ring there.[75] William Windham was also a supporter, but he rarely attended. A bill introduced in the Commons to ban the sport in 1802 was defeated, chiefly as a result of Windham's opposition. The sport was finally brought to an end in 1835.[76] Two forms of entertainment popular with eighteenth-century gentlemen, and the downfall of some, were horse-racing and gambling. Neither of these seemed to have held any interest for Coke. A story, probably recounted by Coke himself to his anonymous biographer, tells of his one and only visit to Newmarket when he was a young man, where he lost not only the money his father had given him for his expenses, but also had to sell his horse to pay his debts. This seems to have cured him of any desire for such amusements.[77]

When he was in London, a small staff would be left in the country and a personal retinue went with the family. For much of this period Coke's London residence was in his mother's house at 19 Hanover Square. The house was on three floors with two attics for servants. As at Holkham, the lowest semi-basement floor was occupied by the servants, with a butler's pantry, housekeeper's room, kitchen, and servants' hall. Above was the entrance hall with porter's chair and three parlours. The drawing-room and other rooms for entertaining were on the next floor and the bedrooms above.[78] There are plenty of references in contemporary diaries to Coke's presence at fashionable dinner-parties and also hosting gatherings himself. In May 1790 Windham dined with Coke and sat next to Burke with Fox next to him 'and had a tolerable share of the conversation'. The principal subjects were 'political, agricultural and artistic'. The following year he dined with Coke again. This time Fox, Burke, Fawkener, 'D. North', Coke's son-in-law Mr Anson, the Duke of Portland, Lord Fitzwilliam, Earl Grey and Lords Tichfield and Petre were there.[79] Coke was a patron of the Royal Academy, where dinners seem to have been gatherings of Whigs. On one occasion when strict protocol in the seating arrangements meant that the Duke of Bedford was placed next to the Duke of Norfolk, Bedford chose to break with etiquette and sit next to his friend Mr Coke. At another dinner Coke was next to Fox, with other Whigs such as the Duke of Bedford, Earl Spencer, the Prince of Wales and William Wilberforce nearby.[80] His noted ability as a conversationalist would have made him a welcome guest. Much time could also be taken up sitting for portraits. He was painted in the first years of the nineteenth century both by Sir Thomas Lawrence and John

Opie, who, along with his wife, the writer Amelia Opie, was part of the Whig circle in both Norfolk and London.

The amount of time spent in London varied considerably from year to year. We know of Coke's attendance at parliament but it is almost impossible to know how long he stayed in town. When the family was away, the servants left at Holkham were put on board wages and the length of time for which these were paid is sometimes recorded in the household accounts. Board wages usually cover the period March to August, although in 1805 they continued into November. In 1802, the steward, Thomas Crick, recorded that the family were away for 28 weeks. Coke was certainly in residence for the sheep shearings in June and the audit in July, as well as for much of the shooting season when he was entertaining guests. Christmas was always spent at Holkham. In his old age, after the death of his brother, Coke spent more time at his childhood home in Derbyshire and he also visited his daughters. His friend and ex-Norwich head master Dr Parr pointed out that his home at Hatton was on the Birmingham to Warwick road, convenient for a visit on his way to or from his daughter Anne at Shugborough, but there is no evidence that this offer was taken up.[81] Jane and their daughters frequently accompanied him to London. A few political wives found London dull. John Parker of Saltram was an MP for Devon and his wife complained that late-night sittings of the House of Commons meant she was left at home with nothing to do.[82] However, the indications are that Jane enjoyed a rich social life while in the capital. Visits to the opera are recorded in her account book. In 1799 the expenses for the journey and stay in London for her and her daughter amounted to £140.[83]

London of the 1780s and 90s was portrayed by the cartoonists as a place of drunken debauchery, where womanising and gambling were commonplace and aristocratic overspending was the vice of the age. The worst offender was the Prince of Wales, who entertained his Whig friends Fox, Sheridan and Burke as well as many others at Carlton House, which he had transformed at great expense. However, there is no record that Coke was ever present on these occasions. Walpole wrote in the 1770s that at Brooks's 'the young men loose five, ten and fifteen thousand pounds of an evening', and described how Fox would arrive at the House of Commons having gambled for more than 24 hours at a stretch.[84] Unlike so many of his contemporaries, it was Coke's good character and immense property that were commented upon.[85] He never frequented the many gaming houses, yet his 'liveliness and playfulness of manners' meant that 'nobody is more entertaining'.[86]

It is surprising that in spite of his friendship with Fox and other leading Whigs, Coke was never offered a ministerial post. Although he was a great talker at social gatherings and to a local or farming audience, he was not a regular speaker in the House of Commons, perhaps a recognition that his political ability was not great. It is possible that he was offered a peerage by Fox as an appropriate way of rewarding a loyal friend lacking the qualities needed for a ministerial post. His uncritical support of Fox was seen by many as a fault: 'He is a bigot to the opinions of the chief of his party.'[87] Perhaps he was seen as little more than a blustering mouthpiece for Fox's views, incapable of original thinking of his own.

Coke first brought his wife, Jane Dutton, to Holkham in 1776, and they remained happily married until she died in Bath in 1800 at the age of 47, but we know very little about her or their family life. Her portraits show a statuesque figure and a strong face with rather long nose and firm chin. She was described as 'clever, well-read and gifted', 'full of dignity and charm' and a 'matchless rider',[88] phrases that could have described almost any genteel lady of the time. Her personal account book for 1789 until her death survives. Her half-yearly allowance was £150, not a very generous sum. Coke's great-aunt, Lady Margaret, had received £400, rising to £500 a year in the 1720s.[89] Much of the account book is taken up with donations, the largest of which was to the Society for the Abolition for the Slave Trade (10 guineas annually), but she also regularly gave small gifts to a 'poor man' and 'a poor girl', as well as the poorhouse in Wighton, the almshouses at Holkham and the spinning school. Unlike her husband she occasionally attended the races. In 1795 a trip to Lichfield races for her and her daughter cost £16. The servants were annually treated to a trip to the theatre and were given Christmas boxes. She frequently bought raffle tickets from her friends at a time when fashionable ladies were organising such events to support French émigrés. The amounts spent on clothes were surprisingly small.[90]

During 1776 Jane gave birth to a stillborn son, a double blow in that no other sons followed. Instead there were two daughters, Jane, in 1777, and Anne, in 1779. A third daughter, Elizabeth, was born in 1795. It was the duty of a mother to supervise the education of her daughters, and Jane Coke took her duties seriously. She corresponded with the noted scholar and Whig headmaster of Norwich School, Samuel Parr, on the subject, and he was a frequent visitor to Holkham. She had read Rousseau's ideas on education, which suggested that noise, laughter and games should be encouraged as approaching nearest to the conditions enjoyed by unspoilt 'noble savage', but confessed to Parr that she could not agree with this philosophy.[91] Parr gave her a copy of his *Discourse on Education*, in which he stressed the need for restraint and correction, noting that the need for coercion would be reduced if reasons were given and 'we attend the disposition of the young person'. Children needed to be praised and 'will be flattered with some idea of [their] own importance when appeals are made to [their] judgement'. In a similar vein he thought that religion should not 'be imposed on young persons as a burdensome task, but recommended as a rational duty'.[92] This advice was at variance with his own teaching methods. He was famed for the severity of his floggings, while at the same time being admired by his pupils, who recognised that he also had 'a heart that soon melts to tenderness'.[93]

Drawing was a necessary and fashionable accomplishment for young ladies. It

Plate 2: *Portrait of Coke's first wife, Jane Dutton (1753-1800) by Thomas Barber. Although we know little of her, the marriage was a happy one in which she played her part as hostess and mother admirably, riding with her guests in the park and taking a close interest in the education of her three daughters.*

was not unusual for well-known artists to be employed to teach them. The Norwich water-colourist John Sell Cotman had a valuable side-line as a teacher, working with, amongst others, the prolific and talented daughters of the Yarmouth banker, Dawson Turner. Gainsborough is said to have taught Jane and Anne Coke drawing, both in London and Norfolk.[94] Elizabeth Coke was well-known for her painting: 'her mother and all her family were great artists'.[95] Although none of their paintings survive, two by Anne (a nest of owls, described by the artist Benjamin Haydon as 'very well done', and a copy of Gainsborough's milk-girl) hung in Coke's sitting-room in 1817.[96] Mrs Powys, who visited Shugborough in 1800, commented on Anne's ability as a painter: 'Every room is ornamented with some of her per-formances', including three full-length paintings of her children 'I think equal to any artist'.[97] A copy of a Titian, done by Jane when she was 15, hung at Holkham and was also much admired. Her talent was certainly passed on to her daughter by her second marriage, Jane Digby, whose sketches of her travels in Syria survive.[98] There are references to the tuning of the harpsichord at Holkham, an instrument that the women all probably played. Two governesses, Miss Attwood and Miss Vrankin, are mentioned in the accounts, although an entry in Jane's account book for 1790 for 'Sundries Miss Vrankin bought for me 15/9d' suggests that the governess might be sent on errands for her employer as well as devoting her time solely to the education of her daughters.[99]

The main concern of the parents of daughters was the securing of a financially stable future for them through appropriate marriages. By the late eighteenth century the choice of a partner was not entirely in the hands of the young person's parents – after all, Coke's own father had disapproved of his choice – but they could certainly hope to be influential by steering their children in suitable directions and introducing them in the right circles. The increasing importance of well-bred 'sensibility' in the late eighteenth century meant that affection was becoming accepted as an important element in choosing a husband, alongside wealth and social standing. Anne was married in September 1794, at the age of 15, to Thomas Anson, then aged 27 and heir to the Shugborough estate in Staffordshire. It is possible that they had met through visits to the Gloucestershire Duttons, although they could equally well have been introduced at London Whig gatherings. In spite of the bride's young age, the marriage was very acceptable to Coke. His elder sister, Lady Hunloke, was staying at Holkham when the engagement was announced and she made sure the information was quickly relayed to London society, as well as the information that Thomas Anson was worth £22,000 a year.[100] Not only that, but Anson's Whig credentials were impeccable. He was already a Member of Parliament at the time of the marriage, and although he rarely spoke he voted regularly in support of Fox, subscribing to Fox's relief fund. The Duke of Sussex described him as a 'true manly, noble, splendid fellow, possessing much of the real English character, sound sense, and although perhaps hurried away a little too much by country sports, has a great deal of good in him'.[101] The two families remained close, frequently visiting each other, Anson frequently Coke's shooting companion. Anne produced four healthy children before she was 20 and seven more thereafter. She was described as 'thin, excitable, energetic, never quiet, constantly getting into quarrels, but always ready to help others'.[102] Dawson Turner's view was slightly different. Commenting on the portrait of her by Barber hanging in the Classical Library at Holkham, he wrote, 'Not a pleasing portrait of a pleasing woman. It has missed her sweet character and has given her an heir [*sic*] of pity [piety?] quite foreign to her.'[103] She lived to 64, a good age for the time.

Jane's first marriage was short-lived. While her choice was Lord Andover, heir to the Earl of Suffolk, Coke was concerned by his lack of wealth. However, they were finally married at Holkham in 1796. As we have seen, the couple were frequent visitors during the shooting season, and when at Holkham, Jane continued to indulge her interest in painting. 'One of her favourite occupations was to sit copying one of the pictures in the house, generally a Poussin, while he [Lord Andover] read Shakespeare aloud to her.'[104] Unlike her sister's marriage, hers was childless and Lord Andover's death as a result of a shooting accident at Holkham in 1800 left her a widow at the age of 23. Over the next few years she frequently visited her father and younger sister. A subscription to Wells Book Club was taken out in her name between 1801 and 1804 and her servants were boarded for up to 12 weeks in several years between 1801 and 1805. In 1806 she married again, this time in London and to a wealthy man, Admiral Sir Henry Digby. His prize-money meant that he could well afford to support a wife and family and they produced four children. They were all frequent visitors to Holkham, often remaining for the entire winter. The four Digby children, along with eleven Ansons, would all be taught together in a large school-room. The oldest daughter, Jane, born in 1808, who was to become notorious for her numerous affairs, finally settling in Syria, married to an Arab Sheik, described her childhood Christmas visits to Holkham as 'wonderful, rumbustious and old-fashioned'.[105] Jane Digby lived to the ripe old age of 86. The close family links of the period are indicated by the fact that Lord Andover's nephew later married the eldest daughter of Coke's sister and brother-in-law, James and Elizabeth Dutton.

In 1800 there was a second death in the family. How long Coke's wife had been ill is not clear. She had gone to Bath in search of a cure, but died there. Windham thought that Lord Andover's fatal shooting accident earlier in the year could have brought on or increased 'her present complaints'. She seems to have been out visiting in the morning 'looking tolerably well', but was taken ill 'with a Seizure' and died during the afternoon.[106] Coke was genuinely devastated by the loss of Jane and commissioned a fine monument by Joseph Nollekens at the phenomenal cost of £3,000, to be erected opposite the entrance to the family vault at Tittleshall. In a characteristic piece, the sculptor depicted a scantily dressed mourning lady. Lady

Plate 3: *Portrait by Thomas Barber of Coke's daughter Anne (1779–1843), who married Thomas Anson of Shugborough, Staffordshire at the age of 15. Frequently complemented on her artistic ability, she was variously described as 'thin and excitable' and 'a pleasing woman with a sweet character'. She produced eleven healthy children and lived until she was 64.*

Plate 4: The monument by the famous sculptor, Joseph Nollekens (1737–1823) erected by Thomas William Coke to his first wife, Jane in Tittleshall church. She was described on her monument as 'Calm and unassuming in the ordinary offices of social life, inflexible and unwearied in the discharge of all its nobler and more arduous duties'. She 'deserved and obtained the love of equals, the respect of inferiors, the attachment of domestics and the gratitude of the poor' and was 'charitable without ostentation'. The style is typical of Nollekens, work and the sorrowing lady in classical pose is said to be a portrait of Coke's eldest daughter, Jane.

Andover, who was said to look very much like her mother, may well have been the model from which he worked.[107] The words that Coke had inscribed on her memorial probably reveal more of what he valued in a woman than anything about Jane herself. 'Calm and unassuming in the ordinary offices of social life inflexible and unwearied in the discharge of all its nobler and more arduous duties' describes a woman who managed the lavish entertaining for which Holkham was famed. She was the one who provided the well-run household which allowed him to shine. Coke's Whiggism still saw the world divided clearly into different social classes. In this stable world Jane 'deserved and obtained the love of equals, the respect of inferiors, the attachment of domestics and the gratitude of the poor'. 'Charitable without ostentation' described the quiet efficiency with which she ran the household and accepted her supporting role in her husband's life. Lord Halifax wrote of the duties of a wife,'the government of your house, family and children is the provenance allotted to your sex',[108] and it would appear that Jane fitted into this role well.

Both Lord Andover and Jane Coke were buried at Tittleshall. Although their funerals were much quieter affairs than that for Coke some 40 years later, the tenantry had to be entertained, a mourning coach was hired for three days and black silk tassels and fringes for the postilions' hats and the altars in church and chapel were purchased. Forty-one mourning suits for the servants were provided. The cost of Jane Coke's funeral was £895 and of Lord Andover's £628. The lack of a son must have disappointed Coke. After Jane's death the son of his brother, Edward, was christened Thomas William (although known as William) in anticipation of his inheriting from his uncle and was a frequent visitor to Holkham.

Elizabeth was only five years old when her mother died. Her upbringing and education then became her father's responsibility, a role for which the landed gentleman of the time was totally unequipped. In later years she told her daughter that no luxuries were allowed in the school-room. She never had a fire to get up by and even on cold days she wore a low dress with short sleeves.[109] It was accepted that 'women are designed by providence to be more domestic, they are endowed with the proper talents and patience to train up their children – men have not the attention that is needed for so great a work.'[110] In such cases children would normally be farmed out to a female relative, and it was Elizabeth's newly widowed elder sister, Lady Andover, who may well have taken on this task during her long spells at Holkham.

It is possible that because she had no mother Elizabeth was closer to her childhood nurse than might otherwise have been the case. In 1822 she wrote, 'My poor old nurse has been so ill that I have been reduced to wait upon myself.'[111] When Lady Andover remarried in 1806 Elizabeth was still only 11, leaving Coke to take on more responsibility for her upbringing. He was certainly very fond of her, describing her in 1809 as 'the comfort of my life' in a letter to Samuel Parr, who took a great interest in her education. Mrs Blackmore, the sister of Coke's late wife, also often came to stay with her children, as did Coke's daughter Anne Anson. Her eldest daughter was about Elizabeth's age, and there would thus have been plenty of young company. We know that when he was in London in 1813, Coke employed an English teacher for Elizabeth, and her lively letters show that she had a good command of language. She was well prepared to take over the running of her father's house when she reached the age of eighteen.

Plate 5: *Portrait by Thomas Barber Elizabeth (1795–1873), Coke's youngest daughter by his first marriage. She was left motherless at the age of five and as soon as she was old enough she took on the duty of hostess for her father. Elizabeth married John Spencer Stanhope of Cannon, Yorkshire in December 1822, shortly after her father took his new wife. We know more of her than her elder sisters because of the survival of her lively correspondence.*

The French Revolution of 1789 had a huge impact of the politics of the day. It split the Whig party, isolating those such as Fox and Coke who persisted in supporting the revolutionaries even as their methods became more brutal, but it made little impression on everyday life at Holkham. Not until war was declared in 1793 were its effects more keenly felt. Firstly, the war meant that agricultural prices would rise and with that came the possibility of higher rents for landlords. Coke's interest in agricultural affairs and the development of his estate farms was therefore given an added impetus and this is discussed in Chapter 4. Secondly, with the threat of invasion, events in France had to be seen in a different light, and even Coke had to accept that for a patriot, the defence of the realm had to come first. In 1794 Pitt steered a Defence Act through parliament which allowed for the setting up of volunteer yeomanry and infantry forces to defend against invasion. Previously the only volunteer force available for internal security was the county militia, commanded by the Lord Lieutenant and primarily intended to keep the peace in case of civil disturbance, but it was not available in time of war. The new yeomanry was made up mainly farmers and their sons, officered by the gentry. They had to provide their own horses, and so long as they agreed to serve outside the county for specific emergencies they were entitled to receive pay for their training days and were supplied with uniforms and weapons. Their role was not to repel invasion, but to maintain the tranquillity of the country should an invasion be attempted. After the passing of the Act in March, Lord Townshend called a meeting in London of Norfolk landowners which was attended by 'nobles, gentry and MPs' to consider raising a subscription for the county's defence. Coke attended and suggested that such a matter should be decided at a County Meeting held in Norfolk and called by the High Sheriff.[112] This was duly called, but the setting up of units of yeomanry for each Hundred, as proposed by Lord Townshend and seconded by Windham, was opposed by Coke. He did not think the war justified it, and his Foxite principles saw a danger in the setting up of armies supported by subscription. The meeting broke up in confusion, but those in favour of collecting subscriptions had raised £10,000 by the end of May. Notably absent from the list of subscribers were Whigs such as Coke and Lord Astley.[113] After Napoleon's failed invasion of Ireland in 1796, which was thwarted by the weather, fear of invasion increased and Coke's lack of concern reduced his popularity within the county and aroused suspicions that he might support the Jacobin element so obvious in Norwich. At a dinner in Norwich in 1797 he found it necessary to state firmly that he was not a republican 'and added that he detested their principles'.[114] Finally with the threat of invasion looming in 1798, he founded the Holkham Yeoman Cavalry. On 6 October1798 the *Norwich Mercury* reported:

> On Wednesday last, two troops of Holkham Volunteers Cavalry commanded by Major Coke received their standard from the hands of Mrs Coke. At eleven in the morning the Troops proceeded to the chapel

where the standard was consecrated by the Rev. Henry Crow in a solemn,
eloquent and impressive discourse pointing out the respective duties of
soldier and citizen.

At noon the troops were drawn up on the south lawn, with Mrs Coke bearing the
standard which Captain Rolfe of Heacham handed to Mr James Bloom, the senior
cornet. This was all followed by speeches and toasts. There was a constant demand
in the country for sergeants from the regulars to come and drill units and Sergeant
John Hilton was brought from London with his family and lodged for 28 weeks to
train the Holkham volunteers. He was provided with a military jacket, waistcoat and
great coat as well as doeskin pantaloons, a helmet and sword. Plated chain epaulets
and a uniform for the Holkham yeomanry were bought for Mr and Mrs Coke, as
well as a helmet, sabre and silk sash for Mr Coke, all at Coke's expense. The
household accounts also record the purchase of 29 helmets, feathers and sword belts.
In December 1798 the yeomanry ball was held at the Fleece to celebrate Nelson's
victory at the Battle of the Nile. The evening was hosted by Coke, now a major, and
Captain Rolfe, and was said to be well attended. Thomas Moore was one of the
tenants who volunteered, and from June 1799 he attended a muster in the park
almost every week. On 6 May the troop dined at the Fleece to celebrate Major Coke's
birthday.[115] By 1800 62 horses and 62 men had been recruited. However, they were
disbanded with the Peace of Amiens in 1802, and on 6 May Thomas Moore, along
with the other volunteers, went to Holkham to deliver up their arms. A dinner was
held and the volunteers resolved to dine annually to celebrate Coke's birthday.

Many other Norfolk gentlemen were raising regiments at this time. In 1794,
Windham raised a volunteer cavalry troop to serve in Norfolk, only to be called
upon in the event of invasion. The cost would not be great as 'I should not propose
their meeting more than once a week; and the expense would be no more, nor so
much as attends their weekly meetings at market. For uniform I would have nothing
but a plain coat, such as they might wear at other times, or no more ornamented
than would make them a little proud of it.'[116] By 1798, however, Windham was
doubting the militias' usefulness. In a secret letter to the Duke of Gloucester he
expressed his fears that should the local militias ever be tried 'the more melancholy
consequences will attend the Trial, not from the men being bad, but from the
Ignorance of the Officers, as they are well aware of their own incapacity to
command they will be diffident of themselves and Confusion will be constant'.
Instead he recommended the recruitment of more regular soldiers.[117] Windham was
also concerned about the vulnerability of the north Norfolk coast near his home at
Felbrigg. In an attempt at coastal defence, the Sea Fencibles had been developed, an
auxiliary corps officered by regular navy officers and recruited from inshore
fishermen and longshoremen, who by joining were protected from being pressed
into the Navy, and a corps was set up at Cromer as well as Kings Lynn.[118]

The peace of 1802 was short-lived, and by May 1803, Britain was again at war with France. The danger of invasion was increased as Napoleon amassed an army across the Channel. For two years his 'Army of England' was gathered at Boulogne. But to invade he would need command of the seas, at least while the army crossed. This was finally denied him after the Battle of Trafalgar in 1805 and the fear of invasion then abated. However, from 1803–5 the country was caught in an invasion fever and more attention was paid to the defence of Norfolk than earlier. Although it was thought that the enemy was unlikely to land anywhere north of Yarmouth, two small batteries which could look over the beach were recommended for Cromer, one at Mundesley, mainly for training Sea Fencibles in the use of canon, and three batteries at Holkham bay as well as Blakeney, Wells and Burnham. There is no evidence that any of these were ever built. Coke, along with others, still thought the threat was exaggerated and showed no interest in re-establishing his militia. The fact that 'In the midst of the general preparations for defence against invasion ... Mr Coke takes no active part' was noted. Instead he went to Shugborough to stay with his daughter. This attitude was 'not easily reconcilable with that *John Bull* feeling for his country' which he had shown five years earlier.[119] However, persuaded by his friends that he should return to reform his volunteers, the Holkham cavalry were re-established at the end of 1803, only to be disbanded again by 1805. The Lord Lieutenants were instructed to reorganise their cavalry into regiments and infantry into battalions. The First Norfolk Cavalry, commanded by a member of the Townshend family, included many local militia, that from Holkham not excepted. The move caused local rivalries and an angry letter from Sir Martin Ffolkes of Hillington, who objected to being 'under the command of an officer of an inferior rank in society'.[120] Several county training sessions were arranged in Yarmouth which were attended by the Holkham contingent. They were inspected in the late spring 1804, when Holkham was placed in the highest category, being deemed to be in 'a great state of forwardness in discipline so as to be fit to be employed in any situation to which Volunteer Corps can be called'.[121] How effective such a force would have been in the face of invasion is doubtful, but at least parading around in a uniform with epaulettes had allowed Coke to feel that he was doing his duty.

Coke has left us very little in the way of personal papers, but it is possible to piece together from other sources a picture of the man. His household accounts show us something of his lifestyle. Total outgoings from these rose from £8,140 in 1791 to £11,256 in 1793. This is a similar figure to the £12,000 a year disposable income that Farington had heard Coke possessed.[122] However, these accounts cover only expenditure at Holkham, not in London, so it is difficult to draw any firm conclusions, except to say Coke does not seem to have been unduly extravagant for the period, when £10,000 a year was thought enough to support an opulent lifestyle. His annual net income in the years up to 1800 was steadily rising, and was in the region of £17,000 to £20,000. As we shall see, Coke had inherited serious financial commitments and to these must be added the expenses of fighting elections. It is clear that Coke's finances were in a precarious state, but as with others of his class, he did not allow this to cramp his style.

Coke's interests were those of an eighteenth-century country gentleman. The development of the park and kitchen garden, as well as the provision of good sporting facilities, were all typical of the period. His concern with agricultural improvement became legendary and will be considered in later chapters. He took less interest in the house and library, while excelling in the degree of hospitality he offered. All this suggests an extrovert personality with little time or patience for academic pursuits, an impression which is confirmed by Farington's conversation with Sir Thomas Lawrence, who had painted his portrait. Lawrence said that Coke had talked incessantly 'and so moved his head about that it was very difficult to obtain a true view of his face. I recommended to him to mention to Mr Coke that he must for a time be steady in sitting or no faithful portrait of him could be taken.'[123] This image of Coke as a great talker who could not sit still fits very well with everything else we know about him. The congenial, expansive and ever-generous host who took pride in his estates and enjoyed the pageantry of playing at soldiering, but who was also apparently a morally upright family man capable of warm affection for his wife and children, is an image entirely in tune with the qualities associated with the eighteenth-century patriot.

A PATRIOTIC FESTIVAL: CELEBRATIONS FOR THE CENTENARY OF THE GLORIOUS REVOLUTION

It is impossible to overestimate the importance of the removal of James II, and the arrival of William of Orange on 5 November 1688, to the Whig party of Hanoverian Britain and its own brand of patriotism. The Bill of Rights, agreed by James's daughter Mary and her husband William of Orange when they ascended the throne in James' place, defined the limits of royal power. Tory ideas of divine right were sorely compromised, if not completely dispelled. For the Whigs, the Bill marked the point when the property owners in parliament protected themselves against future attacks on their liberty and tipped the balance of power away from the monarchy. While the removal of the Catholic James II had in fact been the work of a coalition of Tories, Whigs, Anglicans and Nonconformists, the Whigs had come to claim it as very much their own and it was they who chose to make the centenary on 5 November 1788, an occasion for the celebration of liberty as championed by them. It was to be a time of great patriotic pride in the cherished and hard-won rights of freedom and justice which should be supported wherever they were seen to be flouted, whether that was in colonies across the Atlantic or nearer home.

This optimistic view had not necessarily always been held. Some saw the establishment of a minor European ruler, admittedly alongside a member of the Stuart line, as a temporary measure. When Queen Anne, the last of the dynasty, died in 1715 to be replaced by a German prince who spoke no English, loyalty was by no means universal, with some of the Whigs faltering secretly in their commitment to the new sovereign. They saw it as the moment for the restoration of the exiled Stuarts. It could be said that ineptitude on the part of James and his allies, rather than loyalty to the new king, ensured the Hanoverians' secure hold of the throne.[1]

The date of the landing of William and Mary at Tor Bay coincided with the anniversary of the detection of the Gunpowder plot and so 5 November had a dual significance to the ruling Protestant elite. The centenary day began in Norfolk with a service of thanksgiving in the cathedral. Norwich then celebrated the centenary with dinners at the Swan, King's Head and Maid's Head inns. At the Maid's Head an 'elegant and well-conducted entertainment' was provided, to which Mr Coke had been one of the subscribers. Toasts were drunk everywhere, including loyal toasts to the crown, to the principals of the revolution, the perpetuity of the British constitution, to members of the nobility and the County Members.[2]

However, these events, splendid as they were, were to be upstaged by celebrations at Holkham. Coke saw this as an opportunity to rebuild his political standing, so he set about planning a massive publicity stunt in the form of a 'fete' at Holkham Hall. Organisation began many months in advance, and invitations were sent out in October. As was Coke's style, he claimed the event was non-political. However, it was clear that it would become a celebration of the Whigs and their brand of 'liberty' and particularly of Coke, that larger-than-life patriot, for whom hospitality was as important a mark of an English gentleman as his independence and belief in justice. Out of politics for four years and with the knowledge that an election could not be far off, Coke was anxious to keep himself in the public eye and flaunt his Whig credentials while displaying the hospitality necessary to win votes. This he achieved to an unparalleled degree.

Firstly it was important that the all the right people received invitations. To do this he used his electioneering contacts across the county to provide lists. All beneficed clergy were asked, and other names were provided. Mr Bullock of Shipdham, who was reported to be the wealthiest clergyman in Norfolk, had long been of the Kimberley (Wodehouse) interest, but felt he had been slighted by Sir John and had declared he would never vote for him again. He was now 'independent' and boasted that he might not vote for anyone. 'He is very fond of being taken notice of by great people' and an invitation might well be enough to win his vote.[3] Several guest lists, revised many times, survive at Holkham with crosses and ticks against names. A separate list shows those who had replied and in what terms. Replies are variously described as 'friendly', 'civil', 'very civil', 'zealous', 'very dry' and 'friendly but odd' – all information that could be useful when the election came. Archdeacon Warburton was 'angry at the lateness of his invitation'. Nearly all those invited were from Norfolk. A few invitations went to London, while two, those to the Revd Mr Willis and Mr and Mrs Wayland at Water Eaton, went to Oxfordshire. Also on the list as were the Duke of Bedford and the Prince of Wales with six attendants.[4] The highlight of the event was to be the fireworks, and in an attempt to reduce the resentment of those not personally invited, a general invitation to view the fireworks from the park was issued, with hospitality to be provided from stalls. In this way, Coke could 'stop the mouth of calumny and create fresh friends and interest'. The final guest list numbered about 900 people (or a fifth of the Norfolk freeholders). Most of the familiar names from the gentry are there, but surprisingly only one tenant (Mr Carr of Massingham). No doubt many of them were among the 5,000 outdoor guests.

Organisation and catering for such large numbers involved careful planning. Offers of the loan of plate and cases of knives and forks arrived, as did promises of game for the dinner. Beds were offered to guests from afar, while others said they would make their own way home after the festivities, or stay nearby with family or friends. The Colhouns, coming from their Breckland estate at Wretham, planned to arrive in time to dine at five and were 'grateful for the opportunity of dressing after the journey'. There was also the vexed question of what to wear. Mrs Colhoun asked whether the ladies were to wear a 'uniform' of buff and blue (the Whig colours, adopted at the time of the American War as they were also the colours of Washington's army): 'We are, as you well know, staunch Whigs in this house and would wish to appear in our proper characters.'[5]

On 23 October an Italian expert in fireworks, Signor Marteneli, began work preparing both the fireworks and the decorations. This involved erecting scaffolding and laying the fireworks, charged and ready to light. 'The poor fellow really appears to be extremely anxious to give satisfaction.'[6] George, Prince of Wales, a supporter of the Whigs against his father, the king, was expected to attend, and so the Prince of Wales's feathers were to be emblazoned above the pediment of the front colonnade. In the event, the prince did not come as his father was taken ill with one of his more serious bouts of madness, from which it was possible he would not recover. George had reached Newmarket on his way to Holkham, but returned to Windsor, believing that the crown, or at least a regency, might be imminent.

The guests began to arrive at about eight o'clock, their way lit by illuminations at the triumphal arch, from where they could see, at a distance, the church also bathed in light. Once they reached the top of the drive and looked down at the house from the obelisk the sight was even more striking, with the Prince of Wales" feathers illuminated over the front of the house in the Whig colours of blue and buff. This blatant expression of Whig support gave the lie to the idea that the event was to be non-political, and it is not surprising that leading local Tories had declined the invitation.

On arrival, the guests' names were announced to the host and hostess, who stood at the head of the stairs within their marble hall reminiscent of a Roman temple. Behind them, over the entrance to the grand saloon and the bust of the hall's creator, the Whig slogan 'Liberty and our cause' was lit up. The theatrical nature of the occasion could not have been lost on anyone. Here was the greatest landowner in Norfolk, and yet an independent commoner, a god within his own domains, surrounded by a thousand supporters, in the centre of his estates from which he drew his wealth and which in turn was enhancing the prosperity of the country. Here was a man contributing to the public good, not only by his labours for agriculture, but through his past parliamentary activities. And the fact that this situation could exist and mark out Britain as so different from France and the court at Versailles was, so the Whigs chose to believe, the result of the Glorious Revolution and their influence in the drafting of the Bill of Rights.

All the leading Whig families of Norfolk were present, as well as Coke's close friend the Duke of Bedford. Several bands provided music for different types of dancing in various rooms, but the highlight of the evening was the massive firework display on the lawn and over the lake, lasting from ten o'clock until midnight.

A suitably lavish supper was served at midnight, followed by the full range of toasts. Typical was that of Marquis Townshend from Raynham, who toasted 'The glorious and immortal memory of William III and strict adherence to the Whig principles that produced the Revolution.' There is no record of how much the festivities cost. The household accounts only record very small sums for 'constable duties and cockades'. About 70 men were recruited to police the event.[7]

No doubt, for many of Coke's guests, it was his generosity and hospitality which convinced them of his patriotic fervour and strengthened their resolve in the Whig cause. When the party broke up shortly before dawn on 6 November, many of the guests were found beds for what remained of the night, but the others left. As the carriages of the weary revellers creaked their way home across the extensive Coke estates of north Norfolk, along roads between the fields worked by Coke's prosperous and influential tenants, their brand of patriotism and Whiggism must have seemed as safe and permanent as the great houses to which they returned. As the fireworks exploded and liberty was toasted, no one there could guess that events less than a year away, across the Channel, would test to breaking-point the Whig interpretation of liberty and lead to Coke himself putting on the uniform of a major and setting up the Holkham yeomanry to defend his country against an enemy claiming the motto 'liberty, fraternity and equality' as its own. The French War would require men such as Coke to rethink their definition of patriotism, and lead to political developments in the following century very different from the politics of Coke's early years.

Chapter Four

COKE, THE PATRIOTIC LANDLORD

Every gentleman farmer must of course be a patriot ... in fact, if there be any remaining patriotism in the nation, it is to be found among that class of men.[1]

One of the first duties of a patriot was the improvement of his estates, seen as a moral obligation by the middle of the eighteenth century. As early as 1699, the Scottish landowner Lord Belhaven had written that following the years of famine, it was incumbent on 'all those whom God has blessed with Estates to Double their Diligence in the Improvement of their grounds ... And I am sure that conquest by Spade and Plough is both more just and of greater continuance than that by sword and bow.'[2] However, it was not until the 1720s that these ideas were more generally accepted. Increasing interest in the classics encouraged by an education system which was coming under the influence of the philosophies of the Enlightenment laid stress on the conquest of nature. The increasing popularity of the Grand Tour, which included a visit to Italy, strengthened interest in the ancient civilisations. The cultivation of the land as described in Virgil's *Georgics* was seen as a symbol of civilisation, and so the subduing of nature to encourage the land to produce ever heavier crops became an accepted role for the owners of great estates.[3] The loss of the American colonies helped to bring home the fact that it was unwise to rely on 'distant territories and monopolies for the prosperity of our commerce; these will vanish like the American dream'. Instead the 'only true support of wealth was THE PLOUGH.'[4] As grain prices rose, bringing with them the possibility of rent increases, at the end of the century and particularly during the Napoleonic Wars, the economic as well as moral and patriotic incentives to improve were there.

By the time Thomas William Coke inherited, the roles of landlord and tenant in the business of commercial farming had become established. It was the duty of the landlord to provide the fixed capital in the form of fields, farm roads and buildings, and the tenant, the working capital, such as the seed, stock and implements for the farm. In times of farming prosperity, when prospective tenants were numerous, the landowner could try and pass on to the tenant some responsibilities such as the hedging of individual fields within the ring-fence created by a new enclosure. Similarly, in times of agricultural depression when tenants were more difficult to find and retain, the landlord might have to take some of these tasks back. The responsibilities of the landlord had been fully understood by Coke's predecessors and were ones that they had taken very seriously.[5]

Not all the agricultural commentators of the period agreed as to the benefits to agriculture of large estates. William Marshall, whose *Rural Economy of Norfolk* was published in 1787 after a period as agent for the Gunton estate of the Harbords (later Lords Suffield), bemoaned the decline in the number of small owners, who had dominated many parishes in the county. 'But, among other evil effects of that inordinate passion for farming, which prevailed some years ago, the decline in the independency of this county is a striking one.'[6] He believed that it was amongst the small owner-occupiers in the east of the county rather than on the large estates of the west that improvements in agriculture began. He studiously avoids any mention of Holkham or any of the other large estates in his description of the farming of the county.

On the other hand, a second commentator on agriculture, Arthur Young, had travelled through Norfolk in 1771, and was already of the opinion that it was in north and west Norfolk on the large tenant farms that the new husbandry was to be found. He was clear what were the elements that made up 'Norfolk husbandry', resulting in the intensive farming of that part of the county. His list included enclosure of open fields and commons, the use of marl and clay (the application of clayey, lime-rich subsoil to light acid land) to improve the texture and fertility of the soil, the abandonment of the fallow year by the use of rotations including turnips and a hay crop, the granting of long leases by landlords to encourage tenants to plan for the long term and the creation of large farms.[7] Of these, the landlord was responsible for enclosure and the laying out of large farms, well-equipped with suitable buildings. Marling would always be the tenant's responsibility, while drainage could be a shared cost. The landlord would dictate the length of the lease and could also stipulate the use of certain rotations to allow for increased grain output without exhausting the soil. However, the actual standard of husbandry was up to the tenant, although the landlord could encourage good farming practice by the example set by his Home farm.

The Holkham estate had been built up from the end of the sixteenth century by the founder of the family fortunes and Queen Elizabeth's Chief Justice, Edward Coke. Born at Burwood Hall in Mileham in central Norfolk in 1550, his first acquisition was Godwick Manor in the neighbouring parish of Tittleshall in 1576. By 1616, the main purchases which made up the mid-Norfolk estate had been made in Tittleshall, Wellingham, Weasenham, Elmham, Billingford, Longham, Sparham and Castle Acre. In 1610 Edward bought his first estate in the north-west of the county, when he acquired one of the Holkham manors. In 1612 his son married the heiress to another, moving into Wheateley Hall, on the site of which Holkham Hall now stands. When the Chief Justice died in 1634 he had spent about £100,000 on

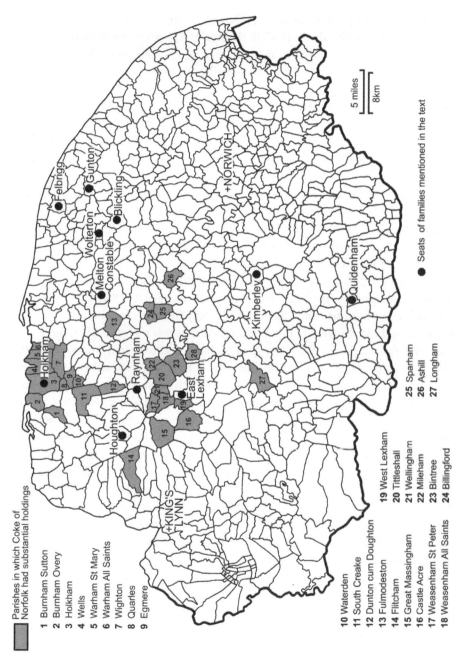

Map 2: Plan to show the extent of the Holkham estate in the early nineteenth century and the seats of families mentioned in the text. It can be seen that there was a concentration of holdings around Holkham, but also a group of parishes to the south around the original family seat at Tittleshall, as well as a few outlying possessions.

purchasing land.[8] Expansion around Holkham continued, with Wighton being added in 1750 and the Burnhams in 1756–7. After the purchase of Warham in 1785, Coke did not seek to expand the estate. Rather he consolidated what he had by buying up small pieces of land within the parishes he already owned. By his death he owned 40,000 acres containing 70 farms across 26 parishes, ranging from the depopulated villages of Quarles, Waterden, Kempstone and Egmere to the populous ones of Tittleshall and Castle Acre. The soils in the two parts of the estate were also very different. The light soils of the north were more suited to sheep and needed careful farming, while the clays of much of the mid-Norfolk estate presented different problems for the improver.

When Thomas William Coke came into his inheritance in 1776 he became owner of some 54 farms on 30,000 acres of the most progressive and well-run estate in one of the most forward-looking parts of the country. North-west Norfolk was already famous for the high standard of its farming, and several of the well-known improving farmers were tenants of Holkham. Coke's great-uncle, Lord Leicester, with the help of his stewards, had been active in increasing the productivity and thus raising the income from rents on the estate by 44 per cent between 1718 and 1759. Farms were increasing in both size and value. In 1727 only 26 farms had been let for over £100, while by 1758 46 were in this category. As this was a period when there was little general rise in prices, the increase must have resulted from improving farm output. The documents show that over this period, land was enclosed and improved by marling, new crops such as turnips and clover-rich grasses were introduced and grown in rotation and new buildings were erected. Maps make it clear that by the time Coke inherited, nearly all the open fields and sheep walks had been enclosed and only small areas of common remained. In all, nearly £50,000 had been spent on repairs and improvements since 1727. Farms were generally let on 21-year leases, which encouraged tenants to invest their own money in their enterprises, and covenants in the leases ensured that their husbandry did not exhaust the light north-west Norfolk soils, but rather increased their fertility. When Lord Leicester died in 1759 his steward, Ralph Cauldwell, who had been appointed in 1742, continued to run the estate for his widow, and rents continued to rise although at a slower rate than previously.[9]

Wenman Coke had set an example to his son at Longford. As we have seen, Arthur Young was impressed by what he saw there. Wenman was an enthusiastic advocate of the use of oxen harnessed rather than yoked – something he shared with Arthur Young and passed on to his son. The Longford farm buildings were also admired. The yards were surrounded by sheds in which the turnips and hay were fed and were divided up for in-wintering different sorts of cattle separately: 'by this means the quantity of manure raised is very great.'[10] Wenman was also responsible for introducing 'Norfolk' husbandry into Derbyshire, involving the elimination of the fallow year and instead growing crops in a rotation which included turnips grown in drills so that they could be regularly hoed to control weeds. He 'sows large quantities, and hand hoes them perfectly, which is a stroke much beyond the farmers in this part of Derbyshire: but the vast benefit this root is to their landlords cannot fail of opening their eyes by degrees'.[11] The importance of marl was also understood by Wenman, who carted lime and farmyard dung into the pits and mixed it with marl before spreading it on the fields. Young concluded his description of Wenman's activities by observing that his 'constant attention' to improvement 'cannot fail of having by degrees a beneficial influence on the husbandry of Derbyshire'.[12] It is clear therefore that young Thomas had been brought up in a tradition of land improvement. His own building activities in Norfolk were very much along the lines that his father had pioneered in Derbyshire. It is also clear that when he arrived at Holkham he found an estate managed in a way with which he was familiar and he would have found it natural to continue to develop it along these lines. It is with the promoting of improved farming that Coke's name is most closely associated and which is considered in this and the following chapter.

Not only did Coke inherit a well-managed and productive estate, but there were also debts. The building of the hall and disastrous investments by his great uncle in the South Sea Company, as well as debts left by Wenman, meant that about £97,000 in loans was outstanding, which required about £4,000 a year in interest payments. As well as this, there were annuities to be paid to his mother (£3,000 a year) and to Mary Coke, the widow of Lord Leicester's son (£2,000 a year). Lord Leicester's will had required that £3,000 a year be paid into a sinking fund towards the eventual paying off of debts.[13] There was therefore a considerable financial incentive for Coke to continue to increase the income from the estate. This became increasingly possible as cereal prices rose at the end of the century. While 'a good understanding between landlord and tenant' was a toast regularly drunk at the sheep shearings and the annual audit, in reality the economic interests of the two were very different. The tenant's aim was to win the maximum production from the soil during his tenancy, while the landlord was more concerned with maintaining the long-term value of his property. Which side gained the upper hand at any point depended on the state of agricultural fortunes generally and thus the demand for farms.

Almost Coke's first action on inheriting was to question the integrity of the steward, Ralph Cauldwell. Lord Leicester's will had stipulated that not only should Cauldwell retain his position as steward for the rest of his life at the generous salary of £300 per year, but also should be a trustee of the settled estate at an annual fee of £200. In 1807 Coke told Thomas Lawrence, while he was sitting for his portrait, that Cauldwell had come to the estate penniless and left a very rich man.[14] He claimed that he took bribes to keep rents down and had defrauded the estate. Legal advice was taken and the claim was dropped, possibly because Coke realised that he could not prove his case. By 1779 Cauldwell had bought the 3,000-acre Hilborough estate in west Norfolk, where he built a 'brick mansion in a well-wooded park'. He continued the tradition of estate improvement which he had begun at Holkham by building three new farmhouses and premises on his estate.[15] He finally left Holkham by the end of 1782, when he was pensioned off with £400 a year. Whether he was guilty is not clear. Certainly his estate accounts were passed as correct by auditors, and Coke made no attempt to alter the rents on Cauldwell's leases. When they expired after 21 years they were not raised more than would have been expected in a period of rising prices. Francis Blaikie, the celebrated Holkham agent, described Cauldwell as 'able',[16] and his advice continued to be sought by other Norfolk land-owners when considering ways of improving their estates.[17] Coke's actions were no doubt prompted by frustration with the terms of his great-uncle and -aunt's wills and the power which Cauldwell seems to have wielded over the old lady. Local gossip noted the power of Mr Cauldwell, and the tenants said that Coke lived at Holkham on 'Mr Cauldwell's sufferance'.[18] Coke may well have been encouraged to harbour suspicions of Cauldwell by his newly appointed 'Auditor General', Richard Gardiner from Derbyshire, who had previously worked for Wenman, both as a political agent and an adviser on estate matters. There was an immediate clash of interest between Gardiner and Cauldwell over who was responsible for what. After surveying the estate, Gardiner had put together a three-year programme for its improvement, but was not allowed to be involved with the audit, which Cauldwell saw as his responsibility. In the event Gardiner himself was dismissed in July 1777.[19] It is clear that the early years of Coke's ownership were fraught with administrative difficulties as he tried to assert his own control over those who had held sway for over 20 years.

Perhaps it is not surprising that having gained control he chose not to appoint a replacement for Cauldwell. He claimed in 1807 to have no land steward, 'but did all that business himself and had done it for many years'.[20] Even if he did not have a steward as such, Coke certainly employed an office staff to keep the accounts. The Home Farm too had its own manager. Consultants were also employed from time to time. The well-established land agent and author of *Hints to Gentlemen of Landed Property*, Nathaniel Kent, was employed from the late 1780s, firstly to conduct a valuation of farms as their leases expired with a view to setting the new rents, and secondly to produce a standard lease, which he published as an appendix to his *General View of the Agriculture of the County of Norfolk* of 1813. Kent's connection with Norfolk went back nearly 30 years. He had begun life as a minor government official in Brussels, but by the late 1760s was advising on land improvement in East Anglia. From 1770 to 1790 he had been employed by William Windham at Felbrigg. For an annual salary of £160 he kept the farm accounts, which he presented to Windham every year in methodically laid out books. He was responsible for overseeing the work of the farm manager and making a survey of the timber.[21] Windham would no doubt have recommended him to Coke. Here again we have an example of the small world of personal recommendations between friends that was so much a part of the political, economic and social life at the time.

In 1816 Coke made his first appointment of a permanent agent to run the estate. Francis Blaikie came to Holkham from Lord Chesterfield's estate in Derbyshire, where he had been bailiff of the Home Farm at Bradley Hall before becoming the estate steward (Colour Plate 8). This is yet another example of Coke's Midlands connections. Nothing is known of Blaikie's background, except that, like many other agents of his generation, he was from a Scottish farming family where he would have gained practical rather than theoretical knowledge. His detractors claimed that he had run away from home and worked as a gardener in early life.[22] Before he came to Holkham he was already in touch with Coke, and had sent him seed from a relation, Andrew Blaikie, who farmed near Inverness.[23] It is possible that, in common with other agents of the time, he would have received some legal training at one of the Scottish universities. If so, he never revealed any academic or professional qualification. Our knowledge of him is gained from the detailed report he wrote on the farms and their tenants when he became agent and his meticulously kept agricultural and estate letter books. His initial report, written on his arrival at Holkham, and his attention to the characters of the tenants as well as the physical state of their farms, show that he realised the importance of the farmers themselves to the improvement and reputation of the estate. Mr Overman, widely recognised as a progressive farmer, was described as 'a very deserving, industrious, attentive and perservering good tenant – a pattern farmer'. Blaikie's lack of respect for book learning is shown in his comments on some of the young men. Mr Heagren of Quarles was well educated but more of a theorist than 'an attentive and practical farmer', while Mr Tuttell Moore of Chalk Farm, Warham, was 'a zealous, indefatigable and practical farmer – and no theorist'. Not surprisingly, Blaikie had no time for Mr Ward of Warham, who was in high spirits: 'He talks of improving, but at present everything is wrong', or Charles Hill, a young man whose farm was a 'disgusting spectacle'. He was a canny enough agent to notice when the estate was suffering by the actions of its farmers. Mr Kendle of Weasenham was taking more crops of grain than the land could stand. After some discussion 'this sensible unprejudiced man' agreed to leave land under grass for longer and so grow wheat every five rather than every four years in the rotation. Mr Brereton's farm at Flitcham could well pay a higher rent and its tenant 'appears to be an eccentric character, but is in reality a shrewd money-getting old-fashioned farmer occupying a very good and very cheap farm'. Blaikie was also looking to the future and the potential of the sons of tenants. Mr Rix, Mr England and Mr Nelson all had sons who seemed to be good farmers in the making. Of young England he wrote 'The son is a very promising and intelligent young man; took him in company around the district and was particularly pleased with him. His information is much beyond his years. He is only 17.'[24] No doubt he was a good pupil and listened politely to Blaikie's often rather forthright opinions.

It may be that Coke decided to employ an agent because with the end of the

Napoleonic war and the likelihood of agricultural depression, economies would be demanded which needed the skill of an agent to implement. Equally, it may well have been that Blaikie's loyalty to Lord Chesterfield meant that he would not leave his employment until the old man died, which he did in 1816. Blaikie certainly thought that Coke had been too extravagant, but it was not until the lean years of the early 1820s that the percentage of income spent on repairs showed a significant decline and then only a very temporary one (Table 1). For much of Blaikie's time in post it remained at between 15 and 20 per cent. He also felt that the estate office needed a great deal of sorting out and summoned the local solicitor, Mr Stokes, who had previously been in charge of local business, to come and conduct 'a regular clearance of this and your office, of the various and numerous unsettled Estate transactions'.[25]

When Blaikie arrived in Norfolk his first task was to reduce the effects of the post-war depression. Coke had inherited not only large areas of north-west Norfolk, but also land in other parts of the country. Like many others amongst the wealthy gentry his assets were widespread. In 1776, he derived an income of nearly £4,000 from small estates in Kent, Somerset, Oxfordshire, Lancashire and London. Unlike many others, however, he chose to concentrate his efforts on his agricultural lands and to sell others to enable him to consolidate his Norfolk holding. In 1786 he sold his property in London, Kent and Somerset in order to buy the Warham estate of just over 3,000 acres for £57,750. Most of the Oxfordshire lands were sold in 1812 to enable the purchase of 1,190 acres at Egmere for £53,400.[26] The inflation of land values between the two purchases is an indication of increasing agricultural profits during the Napoleonic Wars. Between 1790 and 1806 Coke sold all his land near Manchester except for a small piece in Reddish, let by Wenman in 1771. Coke seems to have taken very little interest in these sales, and it was not until 1823 that Blaikie was made aware of their existence when he received a query about the mineral rights on one of them. As Coke's recollection of the terms of the sale was extremely vague Blaikie decided to go to Manchester and see the solicitor who dealt with Coke's Midland affairs to establish the situation. There he found a whole list of estates that had been sold, often with carelessly drafted conveyances which horrified Blaikie. The mineral rights had been sold with all but one of the holdings, all of which had since been found to be rich in coal. Sadly, Blaikie wrote to Coke, 'I wish I had been here 20 years ago, I am even now not without hope of picking up a little remnant.'[27] Only at Pendlebury did Coke still hold the mineral rights on two small pieces of land, but the quantity of coal mined was not as much as expected and Coke received only £300 annually, and this only for 21 years.[28] It seems that Coke, in his single-minded focus on agriculture and his Norfolk estates, had been remarkably casual in his disposal of valuable industrial assets. Other landowners such as the Duke of Northumberland and the Marquis of Stafford had financed their agricultural improvements from coal-mining profits, but Coke had simply thrown this possibility away. Whatever his skills as an agriculturalist, they did not extend to what he saw as less glamorous business transactions.

A further non-agricultural asset owned by Coke was Dungeness lighthouse. The lease of the lighthouse had come into the Coke's possession through Lady Margaret, on her marriage to Coke's great-uncle. It was leased from the crown for £6 13s 4d a year and the leaser was responsible for maintaining the light and undertaking any work required by Trinity House. In return the holder could charge a penny per ton on passing shipping. As trade increased in the eighteenth century, the profits from the lighthouse increased. In the 10-year period 1778–87 the average annual profit to Coke was £3,108, while by 1818–27 it had risen to £6,567. A new lease was negotiated in 1827 which involved an increase in rent from the crown to between £2,000 and £3,000, still leaving a sizeable profit of about £2,000. Very little effort was needed on the part of the Holkham estate to earn this money. A salary of £40 a year was paid to the lighthouse keeper, as well as two cheeses 'as usual'. Oil for the light cost about £200 most years and repairs were usually about £600. In 1790, Trinity House had demanded that a new lighthouse be built, which was designed by Samuel Wyatt and built in 1791 at a cost that could easily be recovered by a year's tolls. There were occasional complaints about the inadequacy of the light, but nothing that required either Coke or Blaikie to visit it. The private ownership of lighthouses was constantly criticised by shipowners, and public opinion was becoming hostile to the idea that individuals should benefit from tolls charged on shipping for the upkeep of lights. The days of private lighthouses were numbered as criticism mounted, and private lighthouses were finally abolished in 1836. Coke received in compensation the generous sum of £20,954 2s 5d. With that the last of Holkham's non-agricultural assets was lost.[29]

Blaikie remained a bachelor. He was a loyal servant of the family and did not confine his concern for its financial affairs to the estate office. When Elizabeth Coke was to marry John Stanhope, he took the liberty of looking into the financial position of the prospective husband. Elizabeth wrote to John, 'I have just had a séance with Blaikie, who for his own satisfaction of course has written down a calculation of your income and charges which he has solemnly given me.'[30] In spite of his careful management of estate funds, he was generous in his private life. A few days later Elizabeth wrote again to John, 'Imagine my surprise last night when he [Blaikie] presented me with a wedding present – a very fine topaz locket set round with diamonds and a beautifully worked gold chain three and a half yards long. When my father spoke to him on the subject, he said "Sir, after my obligation to Lord Chesterfield's family and your own, I could do no less. You have always known that money is no object to me." It really is a magnificent trait, but quite consistent with his extraordinary character.'[31] His dealings with the servant class, however, could be less generous. He was described by Mary Humphries, who felt that she had been wrongfully treated by the Cokes, as 'insensible of domestic comforts and capable of coolly and deliberately recommending the complete ruin of a whole family'.[32] He may have claimed to have no personal interest in money, but when he retired from Holkham in 1832 he had accumulated land in Wells to the value of £1,200, which he sold to Coke. John Stanhope continued to take advice from Blaikie on rents and land prices after his retirement to Scotland. Blaikie's position was taken by William Baker, who had been trained by Blaikie in the office where he had been a clerk since 1821. His salary of £300, which was soon raised to £400, was £200 less than that commanded by Blaikie and is an indication of the different esteem in which each was held.

The first step in any programme of estate improvement had to be the enclosure of land. Although open field by-laws were not inflexible, experimentation was far more difficult in open than enclosed fields and so enclosure was almost a prerequisite of intensive farming. The progressive farmer would want to pasture his animals either on improved permanent grass or leys sown with improved grasses as part of a rotation, rather than poor, overstocked commons. For this reason enclosed farms commanded a much higher (often double) rent than open fields. Thus the rearrangement of the open fields into compact individual holdings was one of the most effective ways for a landlord to increase his rents. The old strips and interspersed holdings had to be removed and replaced by consolidated farms concentrated within a ring fence, and the progress of this process at Holkham can be followed on a series of estate maps. It is clear that in the 1720s many of the estate's farms contained large areas of sheep walks and 'brecks', or land occasionally ploughed up for crops. Intermixed holdings where strips of land were held under different tenancies were also common.[33] However, even by this date changes were taking place. Tenants were being reimbursed for the purchase of hedging and 'bank hurdles' to provide protection against livestock damage while the hedge was becoming established. The audit books record payments for the digging of ditches and laying of hedges at South Creake, Sparham, Massingham and Tittleshall. The most ambitious scheme was that at Godwick, where, in 1726, 117,250 hawthorn hedgerow seedlings were planted and 1,062 rods of new ditches were dug.[34] By the time of the first estate survey carried out for Thomas William Coke by John Dugmore in 1778, small areas of open fields survived in Weasenham, Massingham, West Lexham, Kempstone, Longham and Billingford.[35] However, this did not always affect the farming. In Kempstone, for instance, areas of open field strips are shown on the map, but the implications were purely legal. Mr Heard was tenant of some of the strips, but owned others intermingled with them. 'Mr Heard with the concurrence of Ralph Cauldwell inclosed the same, destroying baulks so no record is left.'[36] Coke soon bought out Heard's property but even before that the farm was run as if it were under one ownership. A similar situation existed in Longham and Massingham. Even where strips survived, they were often held in blocks by individual tenants. Common land as distinct from open strips was to be found in

Map 3: The parish of Burnham Sutton in 1791 and 1840. It can be seen that in 1781 much of the parish was held in strips, while by 1840 a grid of large fields had been created. (Holkham MS E/G4 & NRO DN/TA 358)

Weasenham, Massingham, Longham and Sparham.[37] The audit books list sums of money being spent on hedges in nearly all the mid-Norfolk parishes, as well as Flitcham, the Burnhams and Quarles to the west, between 1776 and 1785. Enclosure could be piecemeal and arbitrary with little in the form of paperwork to record what was happening. A survey of 1786 explains how a field in Castle Acre 'was principally belonging to Thomas William Coke but by private exchange between tenants Mr Coke's Land was given to Mr Hagan for lands of his lying dispersed in field no 344, the boundaries of which had been chiefly destroyed by Mr Coke's tenant'.[38] Maps of 1781 show Coke's holdings in Burnham Sutton intermixed with others and at Fulmodestone he owned 557 acres of intermixed land.[39] By 1794 there had been a considerable degree of exchange, which meant that Coke's holding in Burnham Sutton had been consolidated.[40] The expiration of leases gave an opportunity for the rearrangement of land and this was frequently carried out. When the small Wright's Farm at Wighton came up for re-leasing in 1780s there was much intermixed land, so exchanges of land were proposed and it was suggested that the farm be divided between its two neighbours. At Lower Farm, Billingford the tenant had to agree to give up land 'for purposes of exchange'. Brake Farm, Billingford was re-leased for only 13 years so that its lease would expire at the same time as other Billingford farms and a rearrangement could then take place.[41]

While enclosure of open fields could often be achieved by mutual agreement and exchanges, the enclosure of commons could be more complicated and require an Enclosure Act. Between 1806 and 1816 the estate spent £7,000 on its share of the cost of enclosures. Usually the enclosed land was divided between already existing farms, but occasionally a new farm was created, as at Dunton, and Grenstein in Mileham parish. Here, after the enclosure of the outlying commons in 1814, fields and new buildings were erected by the estate. The drainage of the bog and marling of the land was undertaken by the tenant of this new farm, created from the out-portions of neighbouring farms.[42] Elsewhere the enclosure of commons was often accompanied by a reorganisation of fields. Straight field roads were laid out between rectangular fields so that they could all be reached by a hard track. New sets of buildings were often built in the middle of the new farm. If, however, the buildings were already established off-centre, perhaps on an existing road, field barns might be built to service the more distant fields.

The main period for this final phase of enclosure was during the Napoleonic Wars, when grain prices, and therefore farming profits, reached their highest. It was at this time that the enclosures shown on Dugmore's map of 1778 and Biederman's of 1781 were rationalised and hedge lines moved.[43] Usually the common land remaining was the poorest heathland on the edge of villages and only at times of high prices was it worth the expense involved in bringing it into production. With the end of the war prices fell and agricultural depression set in. Enclosures that had only recently been agreed proved uneconomic to fence, plough and manure. Two commons and an area of waste in the north of the parish of Longham were enclosed in 1813, just before the end of the war. Fields were reorganised and a new house and farm buildings erected on a new site at a cost to the estate of £2,338.[44] However, much of the work involved in bringing the old heath into cultivation fell to the tenant. Blaikie visited Longham in 1816 and reported that 'much money has been expended by the tenant on the improvement of this farm, with a very far distant prospect of return and a very large sum is still needed to complete the projected improvements.' The tenant, Mr Hastings, was a 'zealous and industrious tenant, but heartbroken by his present undertaking'.[45] In the event, Coke agreed to pay the cost of much of the marling and fencing – something he would not have had to do if grain prices had remained at wartime levels. It can be seen that land improvement could be very much a joint undertaking between landlord and tenant in good times, but had to be taken

DATE	RENT (£s)	REPAIRS (£s)	%	DATE	RENT (£s)	REPAIRS (£s)	%
1776	12,332	314	2	1810	23,267	2,836	12
1777	12,769	1,195	9	1811	24,203	4,956	20
1778	12,979	527	4	1812	28,253	4,994	17
1779	13,270	1,195	9	1813	27,736	5,461	20
1780	13,188	735	5	1814	30,008	4,535	15
1781	13,755	930	6	1815	30,833	6,156	20
1782	14,846	1,649	11	1816	31,050	5,815	18
1783	14,862	3,526	23	1817	31,037	5,078	16
1784	14,920	1,715	11	1818	31,227	4,978	16
1785	16,947	1,935	11	1819	31,651	3,578	11
1786	17,227	1,968	11	1820	31,948	4,220	13
1787	17,333	1,638	9	1821	32,186	2,794	8
1788	17,896	1,680	9	1822	31,680	4,280	13
1789	18,309	1,100	6	1823	30,990	5,505	17
1790	18,641	1,392	7	1824	30,950	4,270	14
1791	18,907	1,890	9	1825	31,091	4,114	13
1792	19,039	2,237	12	1826	31,139	5,371	17
1793	19,514	6,036	31	1827	31,296	5,304	17
1794	19,655	3,097	16	1828	31,469	4,350	14
1795	19,763	4,835	24	1829	31,656	5,224	17
1796	20,184	3,160	15	1830	31,784	3,456	11
1797	20,244	7,084	35	1831	31,790	4,503	14
1798	20,353	3,061	15	1832	31,861	2,957	9
1799	20,373	3,141	15	1833	32,059	3,880	12
1800	20,428	2,447	12	1834	32,265	3,531	11
1801	20,507	5,712	28	1835	32,477	5,179	16
1802	20,823	4,131	20	1836	32,528	7,310	22
1803	20,880	5,959	28	1837	32,624	5,266	16
1804	21,324	3,396	16	1838	33,168	4,836	14
1805	21,322	3,078	14	1839	34,176	5,662	16
1806	21,404	2,774	13	1840	35,000	4,375	12
1807	21,839	2,080	9	1841	35,299	4,093	11
1808	22,100	3,010	13	1842	35,299	2,969	8
1809	23,188	3,338	14				

Table 1: Rental value of the Holkham estate and the amount spent on buildings and improvements, 1776–1842. The percentage of rental spent on improvements is shown in the final column.

back by the landlord when prices fell. By 1820 most of this work of land reorganisation had been completed. Where strips owned by tenants were intermingled with those tenanted by them they had mostly been bought out and the process of consolidation, begun much earlier in the eighteenth century, was nearly finished.

A further work of land improvement carried out by several generations of owners at Holkham was the drainage of the coastal marshes. The old coastline had followed the line of the present coastal road, and the first bank, which reclaimed land to the north-west, was built in the seventeenth century. In 1720 a further bank was built to the east and gradually more drainage took place northwards. Once the land was drained, the seaward sand dunes had to be stabilised. In March 1817 a team was paid for 10 days' work carrying faggots down to the sea sand and bringing back reeds for thatching to the carpenter's yard.[46] Marram grass and trees were then planted to hold the sand away from the newly reclaimed the farmland. Elsewhere in Norfolk, in the marshlands between Kings Lynn and Wisbech, drainage had also been carried on from the 1770s, particularly by Count Bentinck in Terrington St Clement. Young reported that he had laid out 'on a scale never practised before' adding 'a thousand acres to his old estate'.[47] The years of peak agricultural prices between 1809 and 1812 saw a speculative boom in land drainage schemes with newcomers buying land at excessively inflated prices. Count Bentinck's heir, Admiral William Bentinck, bought land and encouraged others, including his distant cousin, Lord William Bentinck, to do the same. Buying at the top of the market, he embarked on extremely expensive drainage schemes and was responsible for a new sea bank at North Lynn to enclose part of the salt marsh and was an important promoter of the scheme to improve drainage by digging the Eau Brink Cut.[48]

Interest in farm-building design on the newly enclosed holdings of the north of England can be traced back to Daniel Garrett and his book on farm design, published in 1749.[49] Garrett's book was followed by William Halfpenny's *Twelve Beautiful Designs for Farm Houses* in 1751, which advocated symmetrical courtyard layouts in both Classical and Gothic styles. As the fashion for farm improvement spread with the increase in farming profits a flood of advice on farm buildings followed. Architects and landlords began to take the form and layout of the farmstead seriously and Wenman Coke and his son were no exception. The progressive farmer expected his buildings to meet certain requirements. Firstly, the barn was of central importance, as it was here that the grain would be flailed out of the straw ready for sale. This meant that a threshing floor between two opposing double doors to provide a draught for winnowing was an essential part of the design. The larger barns often contained more than one threshing floor, allowing for several teams of threshers to work at once. In 1792 Coke wrote to Robert Beatson on the question of the capacity of barns. A Scottish ex-soldier who had turned to agricultural writing, Beatson was to be responsible for the section of farm buildings in the Board of Agriculture's *Communications* published in 1797. His reply to Coke was that a barn with an internal size of 57 feet by 24 feet and 19½ feet high would hold 20 to 22 lasts thrashed out. Based on this figure Coke reckoned that the much larger barn at Longlands should hold 69 lasts.[50] Norfolk farmers were criticised by Nathaniel Kent for keeping a quantity of their unthreshed crop in the barn rather than stacking it outside. This led to the unnecessary expense of building large barns. 'What is most recommended is stacking, which ought to be done much more than it is.'[51] The second requirement was for manure. The corn harvest relied on manure to keep the land productive and so the farmer needed buildings to ensure that good quality manure was produced and conserved. The system required that during the winter cattle would be kept in strawed yards where manure built up. Thirdly, the arrangement of the buildings should allow for the efficient movement in and out of the premises of foodstuffs such as turnips, the finished product such as grain, the straw for bedding and the manure for fields. The main components of the farmstead were therefore the barn, the stables, the livestock sheds and the manure collecting yards, and much thought went into the arrangement of the buildings to reduce the amount of cartage involved.

It is difficult to know what farm buildings were like when Coke inherited in 1776. Dugmore's and Biederman's maps show the layout of buildings shortly after Coke took over.[52] Already several farmsteads, such as that at Peterstone and Crabbs Castle, both near Holkham, were laid out around courtyards, while others presented a rather more scattered appearance. Most had large barns, but otherwise the layout of buildings suggests that they had been built at different times, to no particular plan and with no obvious enclosed cattle yard.[53] No description survives until an estate survey undertaken by Nathaniel Kent, which covered 19 farms as their leases expired between 1789 and 1793.[54] A similar survey was made of the farms in the coastal parish of Warham adjacent to Holkham between 1780 and 1790, soon after they were purchased from the Turner estate.[55] All but one farm had more than one barn, and two had as many as four, all had one stable and several more than one. However, livestock accommodation was much less common and there is no mention of cattle sheds. Some of the buildings were described as 'dilapidated', and two sets of farm premises in the central part of the estate around Tittleshall were described as 'old and out of repair'. Clearly there was still much to be done to bring buildings up to standard and made suitable for the type of intensive mixed farming which Coke was advocating.

Visitors to Holkham were always struck by the scale of the building that was being carried out. One of the earliest descriptions of Coke's estate is that by Arthur Young, published in 1784. He was impressed by the 'new-built farm-houses, with barns and offices, substantially of brick and tile, in as complete a style as can be imagined. There is no article that ornaments a country more than this ... and tells the traveller, in a language too expressive to be misunderstood, *we approach the residence of a man who feels for others as well as for himself.*'[56] No doubt this is exactly the message that Coke wanted his visitors to receive, and in this he was typical of many other landowners of the time. Some even attempted to impart a political message. The eleventh Duke of Norfolk for instance lined the road between Penrith and his seat at Greystoke Castle with grand castellated farms which he named after battles in the American War in which the British had been defeated. His visitors were left in no doubt as to his Whig credentials.[57]

Plate 6: The only complete set of farm buildings for tenants designed by Samuel Wyatt for Thomas William Coke was at Leicester Square Farm, South Creake. The cost to the estate was £3,500 paid between 1793–1801. The aerial photograph shows the original layout with a substantial square house built slightly distanced from the farmyard beyond a low wall and curving service wings. Although the house itself is of a style worthy of the minor gentry, its proximity to the farmyard is an indication that it was the home of a working farmer. The barn – a scaled down version of the great barn at Holkham – is behind the house with two opposing wings containing stables and cattle sheds on either side, originally with matching corner pavilions. Those nearest the barn contained feed stores below and granaries and haylofts above, conveniently placed near the animal sheds and yards. The pavilions near the house were shepherd's and bailliff's cottages. The yard was not subdivided into separate cattle yards until the 1870s when the individual feeding of groups of animals became more important. Although this set of buildings is architect-designed it is similar to many other planned farms with large barns built around Holkham during Coke's tenure of the estate.

When Young visited Holkham again, 20 years later, his opinion had not changed. Houses and barns were built in a 'superior style'. A new farm had been built on recently enclosed land at what is now Leicester Square Farm, South Creake, which consisted of an 'enormous barn, with stables, cattle sheds, hog sties, shepherd's and bailiff's houses surrounding a large quadrangular yard, likewise in a style of expense rarely met with'. Young quotes a figure of £100,000 as having been spent by Coke on buildings.[58] How this figure was arrived at is unclear. The headings in the audit books include both new building and repairs, and for the years 1774–1804 the figure amounted to about £80,000. Set against this is the fact that materials were often provided by the estate office. £100,000 might well therefore have been an informed guess given to Young by Coke. It is certainly a large sum and represents in some years over 20 per cent of estate income (Table 1). Again we see Coke courting publicity through the immense sums of money he was able to spend and the highly visible results of that expenditure, which would place him amongst the most exemplary landlords of his age.

However, this degree of exhibitionism was not approved of by Francis Blaikie. In his survey of the estate undertaken when he arrived in 1816, he described several of the new sets of buildings in phrases such as 'undergoing injudicious and expensive alteration', or 'too expensive and extensive for the extent or the rent of the farm'.[59] Later he wrote:

> Mr Coke's tenants are much in the habit of erecting unnecessary buildings ... [which] are not only attended with uncalled for expense to the landlord in the first instance, but entail a lasting encumbrance on the estate. For every particle of building not absolutely wanted is an encumbrance to the estate and a deterioration of the property. These remarks apply more immediately to Mr Coke's estates than to any other in the kingdom. It would be greatly to the advantage of the tenant as well as to the landlord, and much to the credit of the former if they would condescend to be guided by the sound advice that I give them as regards the buildings on their farms.[60]

These remarks are interesting for several reasons. They suggest that new buildings were erected as the result of tenant rather than landlord initiatives, and that once a scheme was proposed Coke was inclined to accept it uncritically. By Coke's death nearly 40 years later, the audit books record that he had spent nearly £233,749 on 'building and repairs'. However, as this did not cover material provided from the estate yard, William Baker, the agent when Coke died, considered that £500,000 was a more realistic sum. 'It appears a very large sum, but when it is considered the Number of Years the Estate was held and the immense buildings which have been created the average will not be thought much beyond what might be expected.'[61] At least half the farms on the estate had been entirely rebuilt with new houses and premises, while all had been considerably altered and extended – an indication of the thoroughness of Coke's improvements.

Clearly one of Coke's most important roles as landlord was to provide the infrastructure in the form of buildings and houses for his tenants. This he did in his customary lavish style, reserving his most impressive structures for the park itself, where they were admired by all who visited. Between 1776 and the arrival of Francis Blaikie in 1816 about 35 major rebuilding projects were undertaken, 16 of which involved the complete rebuilding of farmhouses and premises, with between £1,500 and £3,000 being recorded for each in the audit books. This money was often paid out to tenants in instalments over several years, indicating that the responsibility for building was the tenant's, having received the go-ahead from Coke. In spite of Coke's desire to impress, his buildings were entirely practical and lacked the Classical and Gothic embellishments to be found on some estates. Built of the local red brick with pan-tiled roofs, they consisted of a large, plain barn with cattle yards to the south, as well as stables and granaries. Many of the barns alongside some of the other buildings still survive, often with date stones proclaiming 'T.W.C.' as the owner and promoter of these improvements. Their design was probably suggested by the tenant, who was then responsible for their erection, collecting bricks from the estate yard and supervising much of the work. However, in this, as in so much else, Coke was building on the work of his predecessors. While it would appear from early estate surveys that there was much building and reorganisation of fields waiting to be undertaken after 1776, most of the initial work had been carried out under Coke's great-uncle and the stewards acting for his great-aunt after her husband's death.

One of the most impressive of these groups of buildings was that built at Waterden, where it seems likely that over £1,332 was spent by 1783.[62] Waterden Farm was all that remained of a medieval village, and this sum included the building of a new farmhouse, presumably replacing the Elizabethan one shown on early maps. The long barn contains much worked masonry which must have come from the medieval church, only part of which stands today. The 1783 improvements involved the heightening and extending of the existing barn to include as many as four threshing floors, which made it the largest on the estate. Cattle yards were built on the south side of one of the barn ranges and stables to the west of the other. Arthur Young visited the farm the following year and wrote: 'Nothing in this style can exceed the buildings that Mr Coke has raised at Waterden. Every convenience to be imagined is thought of, and the offices so perfectly well arranged as to answer the great object to prevent waste and save labour.'[63] Even Blaikie was impressed when he described them in 1816 as 'perhaps the finest set of premises in Great Britain'.[64]

Plans drawn of many of the farms in 1828 show the new steadings.[65] At Chalk Farm, Warham, a fine E-plan cattle yard stood beside a second yard behind the

Plate 7: *The buildings at Waterden were regarded by both Arthur Young and Francis Blaikie as amongst the finest in Britain. This view was based on the large area of barn space with at least three threshing floors available, and the provision for cattle in warm, south-facing yards. The cattle yards in the foreground were not built until the 1880s, but the block behind dates to the 1780s. The group consists of an L-plan barn with cattle yards facing towards the camera. The end wings are original and the central dividing one is later. Behind and conveniently placed for the accessing straw and chaff produced in the threshing barn is a stable block to the left. The house beyond is late nineteenth century.*

house, while at Doughton, an already impressive group in 1780, new offices and new barn 50 feet long were built in 1811. The house was also improved, all costing the estate £454. A similarly extensive range was built for Mr Beck at West Lexham, beginning in 1803.[66] At the end of the Napoleonic War new buildings at Panworth Hall, Ashill, were erected on the recently enclosed common on a courtyard plan.

Only between 1818 and 1836 was a person described as an 'architect' permanently employed in the estate office. Two names are mentioned in the audit books, Mr Emerson and Mr Savage. They received salaries of £100 to £120 a year,

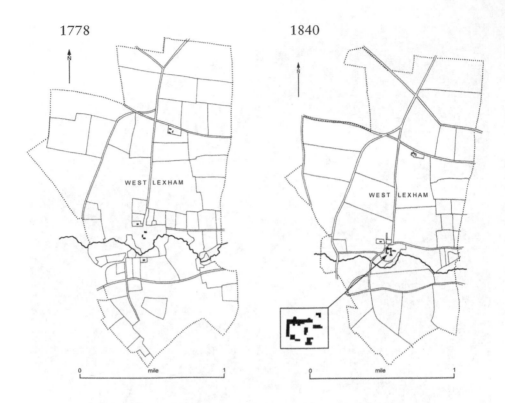

Map 4: The parish of West Lexham in 1778 and 1840. Although only a few open fields survived until 1778, there was a certain amount of field enlargement and re-arrangement by 1840. Like many other farms on the estate, the buildings at West Lexham were rebuilt during the Napoleonic Wars. The enlarged insert shows the long barn with a wheel house for working a threshing machine on the north side, as well as a sheltered yard to the south and ranges of cattle sheds or stables to the west. The impressive house was to the south, overlooking the river. A field barn to the north served the north end of the farm. (Holkham MS E/G1 & NRO DN/TA 28)

less than half that of the agent and more comparable with that of the clerk, and they may have been little more than draughtsmen who drew up plans and then supervised the work on site.

This extensive building programme is entirely what we would expect from Coke's expansive and exhibitionist character, but it went counter to the advice of the more cautious of the land agents of the day. Nathaniel Kent in his book of 1775 recommended economy. Accommodation for animals could often be supplied in lean-tos (cheaper than free-standing buildings). Moreover, he suggested 'barns, which are very expensive, can often be contracted'.[67] When he wrote his survey of

Norfolk in the years immediately after working at Holkham he wrote, 'Farm buildings in this county are upon a very respectable footing, but in my opinion are on too large a scale.' [68]

This almost total rebuilding of tenant farms in a plain style and on a functional layout is typical of some of the larger estates at the time. In east Yorkshire, Sir Christopher Sykes designed and rebuilt at least 14 farms around Sledmere, all on a simple and classically symmetrical courtyard layout with a minimum of architectural embellishments. On the Marquis of Stafford's extensive Shropshire and Staffordshire estates 37 farms were rebuilt to a similarly practical design, although here an estate architect, John Smith, was employed. The Duke of Northumberland employed David Stephenson from Newcastle, and Robert Salmon was 'architect and mechanist' to Coke's friend and fellow agricultural improver, the Duke of Bedford.[69]

Nor was Coke's interest unusual amongst other Norfolk landowners. Arthur Young acknowledges that he was only the 'fairest where many are fair'.[70] According to Kent's accounts for Windham during the 1780s, Windham was spending between a tenth and a twentieth of the income from his rents on buildings and repairs, as well as smaller sums on 'marling, fencing and productive improvements on the estate'.[71]

Only for about 10 years from the late 1790s when he employed the country house architect Samuel Wyatt do we find Coke using a nationally admired figure to work on estate buildings in the park and on several farms. As well as rebuilding many of the lodges, creating the new kitchen gardens and converting the horse engine house adjacent to the hall which had provided the power for lifting water from the well into an octagonal game larder, he was also responsible for farmhouses and buildings on the estate. In choosing Wyatt, Coke was employing one of the first architects of note to give thought to innovative and practical solutions to the design of farm buildings. He experimented with both layout and building materials, and this can be seen in his work at Holkham. His relatives were land agents and farmers as well as architects, and he began his career in his father's architectural office at Lichfield. His friends included industrialists and engineers, and he was always interested in the use of new building materials such as cast iron and the application of steam and water power for agriculture. How Wyatt was brought to Coke's attention is unclear, but it may well have been through his Midlands and Derbyshire connections. By the 1790s Wyatt was working at Holkham while at the same time he was responsible for remodelling Shugborough Hall and building new farm premises there for Coke's son-in-law, Thomas Anson.[72]

Wyatt was probably responsible for rebuilding or considerably improving eight farmhouses and at least three sets of farm premises at Holkham.[73] The value of well-to-do, intelligent tenants with the capital necessary both to improve the land and to buy the stock needed to keep the land in good condition was not lost on Coke. To attract such tenants, Coke built good houses on a par with gentlemen's residences, some of the finest of which were designed by Wyatt. Built of Holkham white brick with slate roofs, rather than the vernacular red brick and pan-tiles, these houses contained many of the architectural features typical of Wyatt's work. Kempstone Lodge Farm, 'combined every requisite of a gentleman's residence' with drawing-, dining- and breakfast-rooms as well as a study, kitchen, housekeeper's room and servants' hall downstairs and five to six bedrooms and dressing rooms upstairs in addition to servants' apartments in the attics (Colour Plates 9 and 10).[74] On a less lavish and more typical scale was Lodge Farm, Castle Acre, where there were two sitting-rooms, kitchens and five or six bedrooms. When Arthur Young visited in 1804 he commented on the farmhouses, which were 'in a style much superior to the houses usually assigned for the residences of tenants; and it gave me great pleasure to find all that I viewed furnished by his [Coke's] farmers in a manner somewhat proportioned to the costliness of their edifices'.[75]

As well as the houses, Wyatt also designed at least four sets of premises with barns of a standard design. On the two tenanted farms of Wicken and Leicester Square the façades of the large red-brick barns are broken by rows of semicircular ventilation panels typical of Wyatt's style. Only at Leicester Square can the complete plan of the original premises still be discerned. The symmetrical layout of a barn with identical wings projecting behind terminating in matching two-storey pavilions provided livestock and domestic accommodation as described by Young. The house was set apart, but connected to the farmstead by curving single-storey ranges containing wash-houses and 'offices'. The link between the farmhouse, and thus the farmer, and his farm was clear. Whilst from his front windows he looked out over the park-like landscape of his home pasture, to the rear were his animals and the manure they produced – a reminder of the source of his wealth. He was not only a gentleman, but also a practical farmer.

Wyatt's two other sets of premises were within the park and built to impress Coke's visitors. One of the ways in which the 'improving' landowner could demonstrate his agricultural interest to both his visitors and his tenants was through the Home Farm. While the rest of the buildings on an estate might be purely functional, those on the Home Farm could be on a grander scale and well-known architects might be employed to build them. The Yorkshire architect John Carr worked on many northern estates, while Sir John Soane and Robert Adam were also active. Designs might include buildings hidden behind Gothic screens as at Hornby and Raby Castles and Constable Burton, all in Yorkshire, and Badminton (Gloucestershire). There was also scope for imaginative layouts, such as the circular plans proposed by Adam.[76] Wyatt was also interested in experimenting with innovative layouts, and had been responsible for several Home Farms before he arrived at Holkham. His first was for Lord Scarsdale, built in 1765, followed by Thorndon (Essex) in 1777 and Sandon (Staffordshire) in 1779. The Home Farm at Doddington (Cheshire) was built in the 1790s, at the same time as he was working at Holkham.[77]

The great barn is the most impressive of Wyatt's farm buildings at Holkham, and

Plate 8: The great barn at Holkham was designed by the country house architect, Samuel Wyatt in the 1790s and admired by those who attended the sheep shearings. It stands today as an icon to the role Thomas William Coke played in publicising improved farming methods. While the house could be seen as a temple to the arts, the barn is a temple to agriculture.

was visited and admired by those attending the sheep shearings. Built between 1784 and 1792, it had a very different layout from the usual plan of cattle sheds to one side of a barn. Instead, the arrangement chosen here was one which Wyatt had used at Doddington and Thorndon, where the barn, as the controlling central point, was in the middle, surrounded by accommodation for cattle. Whilst the cattle sheds have gone, the fine white-brick building with its elegant pairs of porches on both sides and connecting lean-tos, originally containing the gearing for a threshing machine as well as stables and loose boxes, still stands.

The other set of buildings within the park which were the centre of Coke's farming operations was at Longlands. Here stood the older barn, where the sheep shearings had originally taken place. Beside this fine long red-brick building with its three threshing floors dating back to Coke's great-uncle's days, Wyatt built in white brick with slate roofs, U-plan ranges of cart-sheds with huge granaries above. Matching two-storey bailliff's and shepherd's houses stood at the corners, while the new farm manger's house was to one side. Hidden from view were red-brick stables and yards behind a curving wall.

While the main responsibility of the landlord was the providing of the infrastructure in which progressive farming could flourish, it was through the lease that he could have some influence on farming practice. Arthur Young had written that a long (21-year) lease was necessary to give the tenant the security he needed to maintain the fertility of the soil. Twenty-one-year leases had been usual on the estate from Coke's great-uncle's time. In 1727 there were already 25 such leases in force and by 1751 there were 32. Husbandry clauses which laid down the type of farming to be pursued were also important, and these too can be traced back to the 1720s. Rather vague clauses forbade the tenant growing more than three grain crops in succession before the land was put down to grass for at least two years. These are very similar to the terms laid down by other East Anglian estates at this time.[78] Gradually, as elsewhere, leases became more detailed, and one of 1751 for John Carr's Massingham farm stated that no more than three crops of corn should be sown before the land was put down to turnips, followed by two years of grass.[79] The granting of long leases continued under Coke and more detailed rotations were stipulated. A six-year rotation was usual, with three corn crops over six years, one of which was undersown with turnips, and two years of grass. However, the wordings in many of the leases of the 1780s varied and there was no consistent form. When Nathaniel Kent undertook a survey of farms and produced his standard lease in 1789, this was to be for 21 years, with strict conditions for the tenant. He was not to plough up permanent pasture without his landlord's permission, he was to adhere to a six-course rotation with a sixth of his land in turnips or vetches to be fed off by sheep, a third in grass which would remain for two years, a sixth in wheat and a sixth in spring-sown grain, probably barley. To ensure the farmer kept enough stock to produce the necessary manure, all his hay was to be fed to his own stock on the farm, and all the manure produced was to be spread on his farm. He was responsible for keeping the buildings, ditches and hedges in repair.[80] By 1815, the six-course rotation had been modified to what came to be regarded as the classic Norfolk four-course rotation, in which turnips were followed by spring-sown barley undersown with grass, which was then followed by wheat. The golden rule of 'no two corn crops in succession' was thus firmly established.[81]

One of Blaikie's first tasks was to write a series of model leases which would be suitable for all the various soil types on the estate and which he set out in a volume entitled 'The General Form of Leases to be granted by TWC Esq on his Estate in the County of Norfolk with Forms of Covenants for Cultivating the Arable lands'.[82] These leases were used for the rest of Coke's life. Four different leases were laid down, A to D, with four variations of type D. Type A stipulated the classic four-course, specifying that the turnips should be hoed twice, to kill weeds, and fed off the field to sheep. The amount of grass and clover seed to be undersown with the barley was stated. The grass was to be mown only once and fed off by cattle or sheep. Type B was a five-course rotation which allowed for the grass to stand for two years. It thus provided for a less intensive grain system and was suited to less fertile soils. Type C allowed for a four- followed by a five-course rotation, and type D included variations on a six-course rotation for the poorest lands. Crops of beans or a fallow year might be included. All these conditions were to prevent the soils becoming exhausted by too many grain crops (Colour Plate 7). It is difficult to know whether the terms of leases were adhered to or how much they influenced the type of farming carried out. A field book for the entire estate kept between 1789 and 1803, which allows a comparison between the pattern of crops grown with that laid down in the leases.[83] It is clear that a great variety of rotations were practised at the end of the eighteenth century, with four, five and six courses being followed. It was presumably to instil some sort of order that the husbandry covenant became more formalised under Blaikie after 1816. However, evidence from John Leeds's farm at Beck Hall in Billingford shows that tidy systems were easier to put on paper than enforce in practice. Two crops of corn grown in succession continued into the 1820s.[84]

When Blaikie visited Beck Hall in 1816, it was occupied 'under the promise of a lease for 21 years' (to commence at Michaelmas 1816), and farmed under a five-course shift. When Blaikie called to discuss the terms of the new lease it was a type C that he wished to impose. Leeds, on reflection, decided to write to Blaikie to persuade him to change his mind and allow him to alternate in turn a four-course and the Kent-type six-course rotation, which allowed half of the arable to be under cereals. Leeds claimed this would let him grow peas, as 'peas when properly cultivated I am an advocate for, and without being permitted to take them in a six-year course they can never be sown to advantage'.[85] Blaikie, however, did not countenance this argument, and Leeds received the full blast of the Scotsman's advice:

> You have by your intelligence, application and perseverance, added a very necessary ingredient and raised the capability of a considerable part of your farm from a five-course to a four-and-five-course alternately – and some parts from a state unworthy of being classed in any course, to a high state of fertility. It is probable that by continued application of the same means you may raise your land to a still higher pitch. Then follows the query, would it be advisable to do so. I answer <u>NO</u>. A spratt <u>may</u> grow into a herring for any I know to the contrary, but I am pretty certain it will not grow into a whale. Be assured your weaker soils were never intended by nature to bear a four-course shift of cropping and four and five alternate will give you sufficient latitude.[86]

The survival in the letter books of numerous applications by tenants to depart from the terms of their leases suggests that they were taken seriously and that they felt permission was required, even for the smallest deviation. However, an analysis of the admittedly spasmodic entries in John Leeds's journal referring to crops grown in the various fields at Beck Hall suggests that the regular rotations were not necessarily always practised. There were two fields where the golden rule of no two corn crops in succession was broken. Wheat was grown two successive years, and in the case of Home Close in four out of five years. No explanation for this is given in the journal, or in the Holkham letter books. There is no suggestion in the journal that Blaikie ever visited Beck Hall to check what was being grown, and, perhaps because it was so far from Holkham, Leeds felt safe to do more or less what he liked.[87]

Neither Kent's nor Blaikie's leases mention the other major improvements which were necessary to increase the productivity of the soils of north-west Norfolk; namely marling and draining, but as it could be argued that neither was a permanent improvement they were regarded as the responsibility of the tenant. The letter books show that Blaikie expected tenants to undertake both and even pointed this out to prospective tenants. A further land improvement which was undertaken by at least two tenants and much encouraged by Coke was the creation of water meadows. The geologist and engineer William Smith, who had worked on water meadows for Coke's friend, the Duke of Bedford, was consulted and a foreman recommended. Much of the work on the Holkham estate was supervised by a Mr Brooks from Gloucestershire and one assumes he was the same person who laid out meadows for Coke's brother-in-law, James Dutton, at Sherborne. The recently restored meadows at Sherborne were certainly originally constructed at this time and so, yet again, we see the importance of family links. The irrigation of riverside meadows by diverting water from the river and feeding it through channels across the meadow during the winter to encourage early growth had been practised on the chalklands of southern England from the seventeenth century, but was not promoted in Norfolk until the end of the eighteenth. Arthur Young mentions several examples in 1804, two of which had been undertaken by well-known progressive tenants on Holkham farms, and also that Mr Blomfield of Billingford 'on the recommendation of his landlord, Mr Coke has irrigation in contemplation'.[88] In 1804, Robert Overman went over to Lexham from Warham to look at William Beck's meadows and was impressed by what he saw.[89] In 1806 Coke informed Beck that he would be bringing a group over to see the water meadows, and Beck replied that his landlord could bring 'as many friends as possible … I will make them as comfortable as in my power'; they would dine at three o'clock. 'I hope, sir, that it may be the means of inducing some of them to follow your example as without that, the improvement must stop short of general practice.'[90] Further, larger schemes were built in Castle Acre, actively encouraged by Coke.[91]

Such projects were very much after Coke's heart. We are told by Arthur Young that in the 1780s he was looking for a way of providing food for his sheep in the spring when the turnips had either been frost damaged or had run out because the crop had been light,[92] and this is exactly what water meadows, by producing a flush of grass before any was available in other fields, were designed to do. Moreover it was the sort of eye-catching and novel idea which would attract publicity and thus Coke's interest. He offered a 'piece of plate to the value of five guineas' through the

Norfolk Agricultural Society 'to such a person as shall convert the greatest area of waste or unimproved meadows in the most complete manner'.[93] Needless to say, the prize was usually won by one of his tenants.

What then was Coke's legacy as a patriotic landlord? He was certainly clear in his own mind what it should be and wrote to Arthur Young in 1812, describing his achievements:

> I am proud to think that I have the best tenantry in the island and the best cultivated estate with not a single farm unlet. All this is due to the confidence that exists between my tenants and myself and holding out encouragement to them by the granting of twenty-one-year leases. Nearly treble the amount of corn is grown on the same number of acres since 1776. The tenants are opulent and happy and ready to negotiate new leases two years before the expiration of the old. They agree increases in rent which keeps it rolling like a snow ball. Britain should be able to become an exporting rather than an importing nation, but to do this gentlemen must give themselves the trouble (with me it is a pleasure) of attending a little more to their own concerns and be more liberal with their tenantry in granting leases.[94]

This letter was written at a time of agricultural prosperity and rising corn prices so perhaps it is not surprising that tenants were doing well and keen to renew their leases. Coke's reputation meant that during his lifetime there was always a waiting-list of prospective tenants, and families, once established on farms, tended to remain for several generations. But as with so much else attributed to the period of the 'agricultural revolution', change was slow and took several generations. The long leases with which Coke has often been credited had certainly existed before his time, although they were standardised under Blaikie. How far they were adhered to, however, is not clear. Enclosure also was well under way when he inherited but it was important to Coke that his status as a progressive landlord should be obvious in the landscape, and much of north-west Norfolk owes its appearance to his activities. Under his management enclosure was completed and new farms laid out. A landscape of large hawthorn-hedged rectangular fields between straight lanes and roads was created. More importantly, he was also responsible for building or extending the farmsteads on the estate. Although most have undergone considerable alteration to suit succeeding generations, many of his houses survive as substantial, square, late Georgian structures faced in Holkham white brick with slate roofs. But these very houses, which could have passed for gentlemen's residences, are a reminder that the success of Holkham's agriculture rested as much on the progressive tenants of capital who were attracted to the estate by its improving owner, as on its famous landlord himself.

Chapter Five

'THE FAIREST WHERE MANY ARE FAIR'[1]:COKE, THE PRACTICAL FARMER

Norfolk's fame for agricultural improvement was already firmly established by the time Thomas William Coke inherited in 1776. 'Turnip' Townshend on the neighbouring Raynham estate, who died in 1738, was well known as a promoter of improvement, particularly the cultivation of roots grown as part of a rotation. An article in the *Gentleman's Magazine* of 1752 made it clear that he had not invented these changes, but simply adopted them. Improved grasses with higher nutritional values than natural meadow hays had been introduced in the middle of the seventeenth century and turnips from about 1700. These new crops, alongside marling and enclosure, had resulted in 'a course of husbandry utterly unlike that practiced a hundred years ago'. This was particularly true on the traditionally sheep-supporting light soils of west Norfolk. As the value of wool had declined and the price of cereals rose, methods of producing grain on light soils had been developed. Crop rotations could best be practised in enclosed fields, so the old sheep walks had been fenced and marled. Indeed, the writer stressed that it was 'more commonly upon our break [newly enclosed sheep walks] we improve'. As a result five times as many acres of wheat and twice as many acres of barley were grown on light lands that had previously only been broken up for cultivation for a few years before being left in grass for long periods. 'The whole country has acquired a more cheerful appearance', new brick farmhouses were being built, and these had been 'one of the most beneficial alterations that has taken place in the county at any time'.[2]

The agricultural commentator Arthur Young's first visit to west Norfolk was in 1771, five years before Coke inherited, and he set out to describe the 'husbandry which has rendered the name of this county so famous in the farming world'. Like the author of the article in the *Gentleman's Magazine* he thought that change had begun early in the century. Young went on to describe the enclosure of sheep walks, the importance of marling and the introduction of turnips, which were to be hand-hoed to keep the land weed-free, accompanied by the planting of new grasses grown in a rotation to provide a break between grain crops. The landlord encouraged the tenant to farm with an eye to long-term profit rather than short-term gain by granting long leases. The creation of large farms meant higher productivity on the light lands where the keeping of large flocks of sheep was necessary to maintain fertility.[3] Young discussed in detail the farming of Mr Carr of Massingham, who was a great advocate of marling and relied on folding sheep on his enclosed fields to eat the turnips off while manuring the field to increase its fertility. Six hundred sheep were needed to fold across 40 acres in a year. Mr Billing of neighbouring Weasenham was a Holkham tenant who had won premiums in London for his carrots. On the Holkham estate between Burnham and Wells Young saw crops 'better than any I have seen since I entered Norfolk', with both wheat and barley being grown.[4] Although he does not specifically say so, some of the land he saw along the coast here may well have been the on Holkham Home Farm, or Hall Farm as it was sometimes described, where barley was the chief crop and dairy cows as well as bullocks and sheep were kept on the marshes.

However, there were those who argued that it was not on the great estates on the light soils of the west but rather on the smaller yeoman farms of the fertile east that new methods were pioneered. We have already seen that this view was supported by William Marshall, who disagreed with Young on this point. More recently, several historians have given their support to Marshall.[5] It has been shown that turnips were grown on the heavy clays of south Norfolk and north Suffolk by the second half of the seventeenth century and in east Norfolk by the 1680s. In these areas the root crop was seen more as a supplementary feed for cattle than as part of the cereal-orientated rotation it became in the lighter lands.[6] The so-called 'agricultural revolution' was a much lengthier process than was once thought, going back long before Thomas William Coke inherited.

Accounts for the Home Farm at Holkham survive from the 1730s and show that at the time Cokes' great-uncle, Thomas Coke, was building the hall, the farm was being managed on the progressive lines described by Young 50 years later. Turnips were being planted and hoeing them cost the farm £25 in 1733, while £52 was spent on marling. The fact that cereals were being produced is demonstrated by the accounts, which record the cost of harvesting as £59 and threshing £44.[7] The suggestion, supported by Coke himself in his later years, that his estate was a barren desert when he inherited is obviously exaggerated.[8]

While it is clear that many developments had already taken place before Coke inherited, there are important improvements with which his name is rightly associated. It was not until 1785 that Arthur Young made a return visit to north-west Norfolk, nine years after Coke took possession and 14 years after he had first reported on the high standard of farming there. Again he praised the farming of the region, which was the best in Norfolk and where 'the fields of every tenant are cultivated like gardens'. In such a region, wrote Young, where all the farming was of a high standard, it was difficult 'for a gentleman of very large property to make a great figure in husbandry'. In this situation, it was the duty of the landlord to 'make those experiments and introduce those improvements, which [the tenants] are either unwilling or unable to attempt'. Young found that this was exactly what Coke was doing, and his experiments at that time included a search for an alternative feed in the spring when the turnips were either finished or had failed and a substitute for clover on land that had become sick from too-frequent sowing.[9] This led Coke to try sainfoin preserved as a hay crop as an alternative feed, and other grasses as a substitute for clover. All Coke's experiments were carried out with enthusiasm, and as one would expect from his character, on a large scale, for, 'it has never been his custom to try anything by a small experiment when a large one would ascertain the question better.' Beginning with 40 acres, he was now growing 400 acres of 'this admirable grass'.[10] Eight years later Coke claimed that as a result of improving the feed, 2,400 sheep were kept in Holkham parish in contrast to the 700 when he took over in 1776.[11] 'He truly loves husbandry, practises it with equal intelligence and success, and is always most liberally ready to make any experiments that promise to be of public benefit.'[12] In 1804 he was corresponding with the Duke of Bedford on the qualities of lucerne as a feed and later he was promoting the use of cockfoot. In 1813 he claimed to have fed 1,400 sheep and 600 lambs on 220 acres from March to June.[13] In an undated letter, the Gloucestershire and Staffordshire landowner George Tollet, who shared Coke's interest in improved grasses, attempted to persuade him to subscribe to a project to support Mr Curtis of the botanical gardens in Brompton in his experiments with grass. 'He already has 121 species of British grasses and several other plants likely to prove of great service to the farmer.' If he were supported in his scientific work his project was likely to become 'a national institution of the highest importance'.[14] Whether anything came of this is not clear, but as we have seen, 20 years later Coke's search for a feed for his sheep in the early spring led him to encourage his tenants to experiment with water meadows.

Coke's early interest in agricultural experiments and correspondence about them is recorded in his *Book of Observations*, with entries from the 1790s. In it are collected items on a wide variety of subjects: sheep and turnip husbandry, notes on yields of cereals, the manures that were used and the crops they followed. The Earl of Thanet, to whom he was related through his great-aunt, wrote in 1811 recommending the use of lime on turnips to keep off fly. 'It is very unpleasant work and I have recommended to those employed in vain to use crepe to guard their eyes.' Coke was already in correspondence with Blaikie, from whom he received a sample of Scottish turnip seed and detailed instructions on how to plant them. The yellow field turnip 'is also a good table vegetable being more palatable and nutritious and not so watery as the Norfolk variety'. Blaikie was also carrying out milking comparisons between various dairy cows on Lord Chesterfield's estate at Bradley and reporting on these to his future employer.[15] Young again commented on Coke's interest in experimenting when he wrote in 1792 after a short visit to Holkham that he was 'highly gratified with the steady attention Mr Coke is in the habit of paying to the plough'.[16]

Coke's main interest, however, was the selective breeding of sheep (Colour Plate 12). Samples of wool from different breeds were sent to Holkham and sometimes inserted with a letter in the *Book of Observations*. Even a piece of a 'golden fleece' arrived, said to have come from a Chinese cargo originating in Tibet. It is still as golden as ever after 200 years pressed within the book. Recipes for solutions for dipping, dressing wounds, preventing scab and destroying lice and ticks were also collected.[17]

The final years of the eighteenth century saw an increasing interest in sheep breeding, particularly famous being the New Leicesters developed by Robert Bakewell of Dishley and the Southdowns, mainly associated with John Ellman of Glynde in Sussex. Both breeds were fast-maturing and thrived when folded on the turnips which were an essential part of the new rotations. The native Norfolk Horn, however, was slow-maturing and long-legged. While suited to the open commons and foldcourses, it was difficult to contain within hurdles and hedged fields. Arthur Young appreciated the importance of sheep to Norfolk farming: upon the flock everything else depended. 'Corn is the production of the fold; so that to subsist as large a number of sheep as is consistent with their well-being, is the aim of every intelligent farmer.'[18] The value put on sheep is reflected in the amount paid to shepherds. In 1814 Leonard Loose was appointed as shepherd at a yearly wage of £75. On top of this he had a cottage valued at £5 and the use of the heads and plucks of slaughtered sheep, valued at £10. He was also allowed one horse, to be kept at Coke's expense. This is a wage that compares favourably with those of the higher ranks of household servants and the junior clerks in the office. However, even a man with highly valued practical skills could not sign his name. The agreement for the terms of his employment was ratified by him with his mark, a rather hesitant cross.[19]

Young was an enthusiastic supporter of the Southdown breed and from 1784 began building up his own flock. At the same time Coke was experimenting with Bakewell's New Leicesters and played a leading role in promoting them. He noted that Norfolks ate a seventh more than Leicesters and Southdowns, but produced no more wool or meat. Not only was he breeding pure Leicesters, but he was crossing them with the native Norfolks, which he noted were far tamer if kept with the improved breeds. In 1793 Young, along with a group of farmers and landowners, went to Holkham to inspect the sheep. Young's report makes it clear that Coke was trying other crosses as well as the Leicester, but at this time thought the Leicester best. The assembled company measured the length and thickness of the neck and leg as well as the girth of a three-year-old Leicester ram, a young ram and a three-year-old animal ready for slaughter, and all agreed that they were fine animals. The Norfolk crosses were also examined and the length of the fleece measured. Even the cautious farmers present admitted that the cross was superior to the pure Norfolk but muttered that they might not sell, while Mr Bevan and Mr Colhoun, two gentlemen with estates in the light soils of the breckland sheep country, were more ready to innovate and take risks. They 'were so struck by the superiority that they both fixed on tups [rams] of the Bakewell breed, agreed for them with Mr Coke, and sent them directly home to their Norfolk flocks'.[20] Here we see Coke engrossed in his subject in a small gathering of fellow sheep enthusiasts and practical farmers studying in detail the merits of various sheep. When Thomas Moore visited Holkham on one occasion to discuss business he was told that Coke was down at the sheep pens,[21] and there are descriptions of him spending all day in his smock sorting the ewes to go with the various tups.

An early visitor to Holkham was fellow sheep enthusiast George Tollet. His first visit was in 1797, when he wrote a hastily scribbled account of what he saw in the front of a notebook. It was not intended for publication, but instead is a candid description of what he saw. The 'maginificent residence' was 'encircled by planta-tions of Mr Coke's raising'. Inside was the 2,000-acre farm, worked 'by Mr Coke in a most superior style under the immediate management of Mr Wright, a most intelligent man'. Mr Wright showed Tollet round and he was greatly impressed by the 'immense concern'. The Southdown and Leicester sheep were kept separately. Experiments to ascertain the relative weights of the two breeds when fed on the same rations were in progress. Tollet favoured the Southdown as their wool was double the value of the Leicester's. Coke had crossed the Leicester and Norfolk Horn and thought the breed improved. However, there were also Norfolks to be seen in the park. The cows and bull were also of good quality.[22]

The late eighteenth century saw a short-lived but intense interest in Spanish Merino sheep, renowned for their fine wool. British breeders were keen to find a way of improving the fleeces of their sheep by crossing them with Merinos, thus reducing reliance on foreign wool. The project was spearheaded by Sir Joseph Banks, wealthy landowner, renowned botanist and president of the Royal Society, who imported Merinos of French stock in 1785. Landowners were encouraged to send him ewes of local breeds to cross with the Merino rams as part of an experimental programme. Leading breeders such as Robert Bakewell, Lord Sheffield, and Sir John Sinclair sent sheep from Leicestershire, Sussex and Caithness respectively. As well as these, Banks received sheep from Lincolnshire, Shetland and Wiltshire to include in his experimental flock, and in 1789 Coke sent him two Norfolk ewes.[23] Sir Joseph Banks's flock, on the edge of Hounslow Heath, was close to Windsor, and Banks was a frequent visitor at the Castle. King George was an enthusiastic supporter of his scientific work and soon expressed a wish to establish his own flock of Merinos in the park. In 1788 the first pure Spanish Merinos were purchased for the king. This royal patronage would in itself stimulate interest in the breed. In order to concentrate on the management of breeding programmes in the royal flock, Banks gave up his own sheep, passing them on to Arthur Young at Bradfield in Suffolk in 1795.

Meanwhile other breeders were taking an interest in Merinos, one of whom was George Tollet, who bought his first ewes and rams in 1803, showing them at the

Holkham sheep shearings that year. Tollet attempted to establish a breeding programme, but admitted to being hampered by the small number of animals from which to select. As interest in the breed increased and requests for stock arrived at Windsor, Banks encouraged the king to organise a public auction. This was held in August 1804, and there John MacArthur bought his foundation flock, which he then shipped to New South Wales, thus establishing the breed there. The auction stimulated public attention and a second auction was planned for 1805. However, Banks was conscious that none of the leading agriculturalists had attended the sale and that they had complained that the prices were too high. Although it was intended that all the sheep would be sold by auction, Banks felt that the patronage of the breed by Coke would 'do more to interest the general public in it than the attention of any other agriculturalist in the country' and so agreed to sell him three ewes privately before the sale.[24]

The king, already subject to the fits of rage which accompanied his increasing bouts of illness, was furious, and for the first time there was an argument between George III and Banks. Whether it was his personal antagonism towards Coke over his attitude to the American War or simply the fact that Banks had gone behind his back that angered him is unclear. The result was that the friendship between the king and Banks was permanently soured, and Banks resigned his management of the flock. Coke's sheep meanwhile lambed, and his Merinos were always shown at subsequent sheep shearings, where the rams were let. By1806 Tollet had a flock of 80 and reported to Coke that he had sold all their wool for 19s 3½d a tod, while his mixed breed had made 11s 2½d and his English only 5s 6d: 'I feel myself justified in addressing you with so much freedom. Indeed amongst agriculturalists there is something like freemasonry – we are all brethren.'[25] Gradually, however, interest in Merinos declined. They were subject to disease and the enthusiasm of the wool merchants was limited. Although the fleece was heavier than that from native breeds, the price paid did not reflect the quality that was claimed for it. By 1810, Coke's praise was becoming muted. Any improvement in the stock based on the small numbers available would be a slow process which Coke, even with his obvious interest in such things, did not have the patience to undertake. The last year in which Merinos were let at the sheep shearings was 1811[26] and the last royal auction was the same year. The future of the breed lay in Australia, where they rapidly came to dominate after their first introduction in 1804.

In spite of Coke's lack of patience with the Merinos, his interest in sheep breeding in general was genuine and the role he played in introducing the new breeds to Norfolk significant. The early Home Farm accounts merely refer to the purchase of 'wether sheep', giving no indication that many were bred on the farm. By 1814, when the account books resume, it is clear that it was a breeding flock of Southdowns which was the mainstay of the Holkham system.[27] Although Coke was shifting his allegiance to Southdowns, his brother-in-law, Lord Sherborne, was still an advocate of Leicesters. By 1793 Coke was beginning experiments crossing Southdowns with Norfolks and by 1798 was advocating them over the Leicesters, as Leicester ewes were frequently barren, the possible result of excessive inbreeding.[28] Many years later the Southdown–Norfolk cross was recognised as a breed of its own and given the name Suffolk; a black-faced sheep which has since been popular worldwide.[29]

Not only was Coke concerned with the improvement of sheep; he also bred cattle. In 1813 one of his tenants, Mr Bloomfield of Warham, wrote to him, after a visit to Holkham, 'I have very great hopes that Norfolk will soon have a stock of cattle equally suited to our soil and the London market as our Southdown sheep.'[30] In reality this was unlikely. The light soils of north-west Norfolk were ideally suited to sheep rather than suckler cows. It was always going to be fattening of bought-in bullocks rather than home-breds which would be the most profitable. The accounts of Coke's predecessors in the 1730s show 'Scots' bullocks being bought, but by 1814 Devons were also kept. Some had been purchased directly from Lord Fortescue in Devon, while a few were home-bred.

There was revival in interest in the use of oxen as draught animals in the late eighteenth century, particularly on the light lands of the Cotswolds, the southern chalklands and north-west Norfolk. They were used mainly on the largest of farms and to supplement horse power at busy times of year. Coke was a great promoter of the use of oxen alongside horses, and in 1784 he had 'twelve oxen in harness for carting, and finds them a considerable saving in comparison with horses'.[31] In this he was following in the footsteps of his father, who, much to Young's approval, had harnessed 16 oxen together. He found that they worked better harnessed than yoked and was the first to try this method. All his ploughing and home-carting had been done using oxen.[32] Coke retained his interest in Devons all his life, although, according to Young, he had abandoned them for farm work by 1804 because of the difficulty of shoeing them and 'the inveterate prejudices of the men against them'.[33] In 1837 he won a prize for a fat beast at Smithfield. In a letter to Coke at this time, Earl Talbot expressed his admiration of the animal: 'The description of your oxen made my mouth water.'(Colour Plate 14).[34]

One of the most important ways in which Coke displayed both his agriculture and his hospitality was the sheep shearings. It is not clear when these actually began, but they were first reported in the local press in 1798. Both the shearings of 1819 and 1821 were claimed to be the forty-third, which would mean they began almost as soon as he inherited the estate, which seems unlikely. His interest in the improved breeds of sheep can be traced back only to 1784, when Arthur Young reported that he had bought some Leicester ewes to which he had put one of Bakewell's tups.[35] As Coke frequently stated that the aim of the sheep shearings was the improvement of the sheep kept in the region, he would not have started them until he himself had begun keeping and breeding from the new breeds. It was not until 1790 that

Longlands became part of the Home Farm, and the buildings there soon became an important focal point for the event. The Great Barn was built between 1790 and 1793, quickly establishing itself as a significant part of the farm tour.

The early events were simply small gatherings of tenants at which their landlord tried to convince them of the advantage of keeping firstly Leicesters and later Southdowns, either as pure-breds or crossing them with their traditional stock. At the same time he would try and sell or let rams to improve their flocks. From these small beginnings the great three-day events of the early nineteenth century grew. His close friend and fellow agricultural improver the Duke of Bedford began a similar event at Woburn at about the same time. As at Holkham, new showpiece buildings were completed in 1797 and the shearings there began shortly afterwards. By 1800 premiums for the introduction of Leicester and Southdown sheep into Bedfordshire as well as for new implements were being offered. The Woburn shearings were at the beginning of June, while those at Holkham were not until the end of the month or early July, which allowed the dedicated to travel from one event on to the other.[36] In 1802 Arthur Young travelled from Woburn to Holkham with Coke to attend the shearings there for the first time. Inevitably he compared one with the other. Two hundred people were entertained to dinner at Holkham that year, 'The dinner better than at Woburn, I think from its vicinity to the sea which gives plenty of fish.' He was disappointed that although Coke was keen to promote his Home Farm and his own sheep as well as some of the achievements of his tenants and 'was personally civil and attentive', he 'took not the slightest opportunity of mentioning me, the Board [of Agriculture of which Young was secretary], my report or anything about it, though the occasion called in reason for it'.[37] In 1804 it was Sir Joseph Banks who accompanied Coke on the three-day journey from Woburn to Holkham. Banks was treated as an honoured guest at Holkham and eagerly participated in the further three days of agricultural events.[38]

Detailed descriptions of the sheep shearings are to be found in the local press. Fortunately the final gathering, in 1821, was written up in detail in a pamphlet of more than a hundred pages by the editor of the *Norwich Mercury* and admirer of Coke, Richard Noverre Bacon. Although the numbers attending increased steadily over the years, the programme varied little. On the first day the company assembled on horseback at 10 o'clock on the lawn in front of the house, where in 1821 Mr Herod of neighbouring South Creake had brought a party of little girls from his flax mill. They were busy spinning and carding the flax, thus demonstrating the value of the crop for providing rural employment 'for nothing is so delightful as happy industry'.[39] The 'numerous cavalcade' of horsemen then moved on to Longlands. On the way the various crops growing in weed-free fields would be pointed out, and then, at Wyatt's fine new white-brick and slated buildings, they dismounted to admire cattle, horses and boars. Implements of new design would be laid out in the extensive cart-lodges under the granaries. In 1821 these included a drill roll for light land and a plough with a drill attached on one side, while in 1807 there was a turnip drill and an 'inclined-plane plough adjusting its depth of furrow by a screw invented by Mr Balls of Saxlingham and made by Thomas Boyce, wheelwright of Langham'.[40] Beside the new buildings at Longlands was the huge old red brick barn where the sheep were being shorn (Colour Plate 11). As the sheep were the main purpose of the visit, the visitors crowded into the building to see Thomas Coke and the Duke of Bedford, along with the other expert breeders present, inspecting closely the quality of the fleece and the conformation of the shorn animals and to hear their comments. The tour then moved on, out of the south gate to the tenanted farms on the south side of the park, where they would admire the fine crops, stock and buildings, and hope to have an opportunity to put their own questions and comments to their ever-obliging host.

The reports give no clear idea of how many people took part in these grand tours, but we know that in 1821 300 people were entertained to dinner on the first day, rising to 700 on the last. The sight of a large body of horsemen riding around the fields and into the various farmyards led by the owner of these spreading acres and greeted by the affluent tenants with bailiff, teamsmen and shepherd in attendance, was something long to be remembered. It is hardly surprising that it made such an impression on the American visitors. The American ambassador, William Rush, who attended the shearings in 1819, recalled how Coke led the 'informal discussion and explanation on everything connected with agriculture in its broadest sense, on his grounds, or at the dinner table, and even more impressively on horseback ... he plays the part of the old English country gentleman as he rides from field to field attended by friends who are also mounted.'[41]

By three o'clock everyone had returned to the hall for dinner. The main tables were laid out in the statue gallery, while others were in the various rooms connecting with it. After dinner, those in the outer rooms came to stand in the gallery to hear the speeches. The first toast in 1821 was to the 'King and the Constitution', and as the king's younger son, the Duke of Sussex, was often present the second would be to the 'Royal Family'. All the important people present would be picked out for a toast to which they replied, praising their host and his farming in glowing terms. Not only would the British aristocracy and leaders of farming be named, but also the foreign visitors. In 1821 this included the French Consul, marked by a toast to the 'prosperity of the agriculture of France'. These events were claimed to be non-political, and certainly the speeches were mainly on agricultural matters, Coke, for instance, extolling the virtues of the Southdown over the Leicester sheep. The Americans were always treated with particular kindness, with Coke noting in his introduction to the toast that 'Everyone knew his early respect for the Americans for their many and independent assertions of their liberties.' It was not until the last shearings of 1821 that politics entered into the speeches. With agricultural depression deepening, Coke and many others felt free to criticise the corruption and waste that they saw in the current Tory administration and called for the repeal of taxes on malt and agricultural horses. The toasts concluded, the company returned to the sheep pens, and Coke would hope for good prices from those who had enjoyed his hospitality in an auction of his Leicester and Southdown rams and ewes.

The second day was taken up with visits to the sheep pens, where the shorn animals were on show. Riding to the north end of the park, the company saw the new cottages being built and in 1821 they admired the new school. Turning back to the Great Barn, they saw fields of wheat and barley on the way. In the sheds surrounding the barn were the prize cattle. The symbolism of the juxtaposition of the barn and the house would not have been lost upon the guests. One of the most famous pictures of Coke is that by Thomas Weaver showing him as a farmer with his shepherds amongst his sheep, possibly at the Great Barn, with the hall clearly visible amongst the trees in the background. The great patriot landlord was not only custodian of a mansion that was, in the words of its creator, 'a temple to the arts' but also of agriculture, of which, in a sense, the fine Classical, architect-designed barn was the temple. The wealth from one allowed for the maintaining of the other. The two were inexorably linked.

Dinner that afternoon was attended by 500 and was followed again by toasts and speeches. Speakers in 1821 included Sir John Sinclair, previously President of the Board of Agriculture, whose work on his Caithness estates was well known, and Robert Owen of New Lanark, whose reply, whilst heaping effusive praise on his host, cast doubt on the present economic system of landlord and tenant and instead expressed a desire to see a more cooperative organisation. The 'Laird of Skene' was also toasted, but his reply, according to Bacon, was 'of some length, but in a very subdued tone'. Dinner was followed by further visits to the sheep pens in the hope of more sales. In 1802 Mr Moore, tenant of one of the Warham farms, went over to the shearings on two separate days, coming home to dine on the first day, but staying on the second along with 200 gentlemen. He 'bought ten Southdown shearlings of Mr Coke's by auction'.[42]

The visits on the third day were somewhat curtailed, as dinner was at two to allow time for the prize-giving. The carcasses of the slaughtered sheep were inspected in the slaughter house. This final dinner was by far the most popular, with 700 guests in 1821, and after it there were more toasts and speeches. Prizes for the stock, for the ploughing matches and for the most useful newly invented implement were awarded, many going to Coke's tenants.

After dinner the local guests drifted away, while some of the 50 or so staying in the house would remain for at least that night. In 1819 William Rush was taken by Coke for a ride to Wells, when conversation was general as well as agricultural. They returned at dusk to find the other house-guests chatting and playing whist. There was supper in the statue gallery at 11 p.m. for those who were still up, and conversation went on until just past midnight.

While some might stay on for a day or so after the shearings, others arrived in advance. George Tollet came a few days early in 1806, having finished shearing his own flock and sold the wool, to have a quiet look round and see how the preparations for the shearings were organised.[43]

The final shearing of 1821 was certainly the largest, and as we have seen the agricultural depression was much discussed. In the following years depression increased and even the well-run Holkham estate was having to reduce rents. This decline in optimism, plus the fact that in 1822 Coke remarried, may explain, why, at the age of 67, Coke felt it was time for a change. Coke's close friend Francis, the fifth Duke of Bedford, had died in 1805 to be succeeded by his brother John. The Woburn shearings had ended in 1813 and John began to lose interest in agriculture, resigning from the Smithfield Club in 1821, so Coke lost a loyal supporter. Although it seems that Coke was a man of bounding energy, perhaps even he was beginning to feel the strain of three days of intensive entertaining and activity.

There was, however, a different view of the event. Arthur Young, who after the tragic death of his daughter, Bobby, took solace in religion and became more of a recluse, was urged by Coke to attend the shearings of 1806. He refused the invitation.

> There is not one feature that would carry a Christian there for pleasure, but a thousand to repel him, and this is so much the case with all public meetings that they are odious. The Norfolk farmers are rich and profligate; coarse oaths and profanities salute the ear at every turn; and the gentlemen and the great, when they are without ladies are too apt to be as bad as the mob and many of them much worse ... much as I love agriculture, I can renounce it with more pleasure than I can partake of it thus contaminated.[44]

An afternoon of drinking toasts could obviously have some less attractive side-effects not mentioned in the more elegiac accounts of the proceedings.

Thomas William Coke would have excelled at the running and hosting of an event such as the sheep shearings. He loved the publicity, which would certainly have enhanced his county standing, both as a politician and a farmer. It ensured that Holkham would be known as one of the most progressive estates in Britain and the owners of estates who might aspire to share that title, such as John Sinclair from Caithness, Lord Spencer from Northampton, Sir John Curwen from Workington and the Duke of Bedford from Woburn, were always amongst the most honoured guests. Whilst the shearings had begun as small events to promote the improved breeds of sheep, by 1821 the object was 'to bring together eminent agriculturalists from all parts of the three kingdoms and from foreign countries for the purpose of facilitating enquiries into and promoting discussion upon the most approved practices of husbandry and diffuse that useful knowledge for the benefit of mankind in general'.[45] In these events we see the patriot where he should be: on his estates amongst his loyal tenantry, promoting that which was his first and foremost duty – increasing the productivity of the soil.

An important part of the programme at the sheep shearings was the visit to the Home Farm, variously described as the Park or Hall Farm. Most estate owners kept a farm in hand, either within or near the park, partly as a source of food for the house. However, increasingly during the eighteenth century it became a place where the landlord could try out new ideas and demonstrate his interest in 'improvement'. The importance and thus high status with which he regarded the activity of farming could be demonstrated in the erection of impressive, often architect-designed farm buildings, and as we have seen, Samuel Wyatt worked on both domestic and farm buildings on the estate. As well as producing food for the household, the home farm was sometimes expected to set an example to the tenantry. It could do this not only by adapting new methods, but also by making a profit. The Duke of Bedford expected his bailiff to produce a 5 per cent profit on capital invested, but this was unusual and hardly ever achieved.[46]

At Holkham the Hall Farm had been taken over as a farm in hand in 1722 and immediately stocked with sheep. Traditional crops such as rye, wheat and barley were grown as well as lucerne, clover, nonesuch and turnips; crops usually associated with the 'improvers'. Labour accounts show turnips being hoed, rape cake being bought as a fertiliser and marl being spread in the 1730s; well before Coke inherited.[47] The household was charged for goods provided from the farm, and the farm paid a rent to the estate. Taking these into account, the farm showed a profit from the 1730s, and was clearly a serious agricultural concern. During the first few years of Coke's management land was added to it. In 1780 the farm contained about 1,300 acres of arable and marsh but by 1797, according to Tollet, it had increased to 2,000 and was thus the largest farm on the estate. In 1823 a further 500 acres were added. It was here that Coke carried out his breeding experiments with sheep and brought the many farming guests who called at all times of year. 'You cannot please me more', Coke wrote to Arthur Young, 'than by recommending intelligent men to come and see my farm. It is from them I gain the little knowledge I have, and derive the satisfaction of communicating improvements amongst my tenantry.'[48] Tollet was one such visitor in 1797. He noted that all the arable crops were a drilled. A hundred acres of wheat were grown and dressed with powdered oilcake as a fertiliser. 'The land is light and naturally very indifferent: but much to the credit of Mr Coke's liberal management abundant crops are got.' Some of the poorer land was kept in sainfoin for 12 years and hay made. Turnips were grown in rotation, but when the ground was too wet for them oats and barley were sown with grass seed, which followed the grain crop instead of turnips and then wheat was dibbled into the grass. This strange and very labour-intensive method was said to prevent the wheat running up to straw and then being laid in bad weather – something that was inclined to happen if wheat was sown after fallow. It is recorded nowhere else and it may be that it was being tried experimentally when Tollet visited. He went on to describe the new farm buildings, paying special attention to Wyatt's new barn 'on an elegant plan', with its yards for 300 cattle.[49]

Not only was the farm well stocked with sheep, cattle and pigs, but it was also well equipped. The size of the operation can be gauged from the 13 wagons, 12 tumbrels, two turnip carts, one scotch cart and one market cart listed in a valuation of 1817. Twenty-seven pairs of harness, 40 plough collars and 50 halters are an indication of the number of horse teams that could be working at any one time. Although the enormous number of 62 horses was kept on the farm, some were for work in the garden and others for the use of the glazier, carpenter, bricklayer and architect. The farm was also mechanised. Two threshing and two winnowing machines operated during the Napoleonic Wars. The total value of farm equipment came to just under £1,000.[50] Threshing machines, first introduced in the 1780s, were found only on the largest of Norfolk farms at this time. Labour was cheap and plentiful in East Anglia, and hand-flailing provided much-needed winter work. The use of machines merely caused unemployment and so an increase in the poor rates paid by farmers. As the problems of unemployment increased with the depression at the end of the Napoleonic Wars threshing machines were being abandoned. Blaikie wrote in 1822: 'Thrashing machines and all other Agricultural implements calculated to abridge manual labour are now very generally falling into disuse.'[51]

The farm accounts show that corn was the most important source of income. In 1814–15 sales amounted to £3,606. Sheep were the second most important and the 2,004 sheep brought in £1,643 from the sale of wool, surplus ewes and lambs. While the keeping of bullocks alongside the sheep was essential for the manuring of land, the 160 neat cattle produced an income of only £512. These were nearly always sold in small batches at Smithfield. The expense of droving and selling seven bullocks amounted to four guineas.[52]

It is very difficult to calculate how much labour the farm employed on a regular basis, as gangs were brought in to perform particular tasks. Thirty-nine men were paid for 18 days in 1815 to hoe turnips and cut wheat at 3s 6d a day. 'Fuller and 18 men and women reaped 18 acres of wheat' at 3s a day. Some men were employed on a weekly basis, and the bill for this for the year 1814–15 was £1,576. There were 11 living-in farm servants hired by the year and their pay amounted to £722. The cost of boarding them at 8s a week was £270 for the year. These men would have been responsible for livestock and thus employed on a more secure basis.

Accounts and valuations for the years 1817 to 1826 survive. They are laid out in such a way as to record all sums of money going in and out of the estate account and thus serve to monitor the honesty of the bailiff, rather than to show profit and loss. They cover the years of depression following the Napoleonic Wars and the economic slump of the early 1820s, and so, perhaps not surprisingly, they show a loss in most years.

The Home Farm, along with all the others on the estate, was described by Blaikie in his initial survey of Coke's farms. The farm buildings were 'magnificent', but those at Longlands were to the extreme south of the farm and therefore not well situated. The soils were varied, naturally inferior, 'but changed by the great effect of superior cultivation'. The farm was 'superb' and all the land 'beautifully clean'. A four-course rotation was usual, although on the poorer soil the land would be left for three or more years in grass before being returned to the rotation. The stock were all well selected and 'highly approved'. Only the pigs allowed for further improvement. Blaikie stressed the importance of experimentation. Different grasses were being tried, as well as the transplanting of turf to form pasture rather than the sowing of seed. This process of turfing, known at the time as 'inoculation', was experiencing a short period of popularity. 'Every feasible plan or scheme of improvement in agriculture has here a fair trial and if successful and proved so, goes forth to the country at large, recommended by actual practice.'[53] This is exactly what Young had advocated 30 years earlier: 'the great object, that ought to actuate the former [the landlord], is to make those experiments, and to introduce those improvements, which the latter [the tenants] are unable or unwilling to attempt.'[54]

All visitors to the Home Farm were agreed that the soil was naturally poor and could be made productive only by careful farming. This always involved an initial dressing with marl, dug from pits across the farms. On the Home Farm 80 to 100 loads per acre were spread after the barley harvest on the new grass sprouting through. The grass was then usually left for at least three years before the land was ploughed, 'which is far better and more durable than ploughing it directly'. The result was that up to 300 acres of high-yielding grain was produced on the farm every year.[55]

Blaikie's agricultural letter books refer to experiments being carried out under his guidance on the Home Farm. He was frequently sent seed and samples of fertiliser to try there. Producers of gypsum and salt wrote to him suggesting that their products might be useful as fertilisers. While Blaikie agreed that gypsum could be useful on wet soils because it absorbed moisture, he reckoned that Holkham was not the place to try out salt as it was too near the sea. He suggested Lord Albermarle, also an 'advocate of improvement', whose estates were well inland. Sometimes a note of weariness is evident in Blaikie's letters. In June 1821, he wrote to an Aberdeenshire farmer who had sent him turnip seed: 'There were a great many trial seeds sown on Mr Coke's farm last year, but the produce from none was so much approved of as the sort normally cultivated at this place.' While it was Blaikie rather than Coke who wrote copious articles in the *Farmer's Journal* and replied at length to the numerous queries raised, there is no doubt that the two worked closely together and that Coke had an eye for detail. In February 1829, Blaikie wrote to Mr Garwood, the tenant of West Lexham: 'Mr Coke has directed me to mention to you that on his ride from East Lexham to Weasenham he observed the hedges of one of your fields had been cut recently and he thought the work not so well executed as is customary with you in your husbandry operation.' The leading stem had been left too high and not notched near the bottom, which would result in strong branches being sent out at the top, leaving the hedge gappy lower down.[56]

During the agricultural distress of the early 1830s, threshing machines, which had been creeping back on to the largest farms, were destroyed and ricks burnt on the Holkham estate as elsewhere. November 1830 saw the beginning of a sustained series of attacks which swept through 150 Norfolk parishes, many of them in the north and west of the county. Arson and machine-breaking were often preceded by letters signed by 'Swing' sent to individual farmers, threatening action if the farmers did not dismantle their machines themselves. The identity of Swing has never been established and the origin of the code-name is obscure. It is unlikely that there was any central planning; rather the riots spread spontaneously from village to village.[57] The week ending 27 November was the worst, with 'scarcely a night passing without an occurence of this deplorable kind'. While rick-burning usually took place at night, machine-breaking could be carried out in broad daylight. The *Norfolk Chronicle* declared that 'firm, prompt and spirited resistance was needed by landed proprietors'. Sir Jacob Astley offered a reward for information which led to the apprehending of those who burned ricks at Briston, near Melton Constable Hall. Later in the week a mob intending to attack the hall was dispersed.[58] Coke delayed his return to London for the autumn session of parliament because of the unsettled state of the county. While many farmers were said to be willing to dismantle threshing machines in order to provide winter work for an impoverished work-force, Henry Abbott, one of Coke's Castle Acre tenants, prepared to meet the mob, and, undeterred, kept his machines regularly at work, but eight men stood guard every night over his stables.[59] Riots at the Burnhams and Sparham on the Holkham estate are recorded in the press, and Thomas Keppel recounted the story of Coke setting out with his chaplain Mr Collyer to face a mob near Burnham. He stunned the rioters into silence by addressing them and then collaring four ring leaders, whom he drove to Walsingham gaol – quite a feat for a man of 77![60] In 1831 Coke gave a donation of a pound to the Launditch Association, whose aim was the protection of agricultural property within the area, and in 1832 he felt it necessary to pay two men for 'watching the house during the family's absence for ten weeks until the 10th July'. References to the apprehension of poachers increased at this time.[61]

After the retirement of Blaikie in 1832 the estate was run by William Baker, who had joined the estate staff in 1821. He continued to use the model leases developed by his predecessor and must have taken on an increasing responsibility for the estate as Coke spent more time at Longford in old age. Although Coke still basked in the glory surrounding his agricultural reputation, he seems to have become rather disillusioned with agricultural progress in his later years. In 1836 he refused to allow the West Norfolk Agricultural Association to add his name to a letter to the High Sheriff of the county asking him to convene a meeting for the purpose of petitioning parliament to grant a committee of enquiry into the state of agriculture. Coke replied that he did not think there was much the government could do as 'supply and demand will always regulate the market'.[62] A final vitriolic letter, copied into the agricultural letter book and dictated to his daughter, Anne Anson, in his eighty-seventh year because his eyesight was failing, shows that he still favoured the eighteenth century 'practical farmer' over the scientifically minded high farmer of the mid-nineteenth century. In it he refused to have anything to do with the newly formed Royal Agricultural Society of England, which he thought would do 'more harm than good' and would not lead to 'improvement in any way'. He regarded 'all attempts to introduce chemistry as an engine of cultivation as a complete fallacy'.[63] Perhaps by now he simply felt that progress was something that he could no longer keep up with and found it difficult to make way for the younger generation.

Earlier, Coke had been active on the national agricultural scene. The 1790s saw considerable interest in parliament, initiated by the Scottish MP and agricultural improver, Sir John Sinclair, for the setting up of a Board of Agriculture. Sinclair envisaged that the Board would collect information from home and abroad on agricultural matters, encourage experiment and publish the results with the aim of promoting improvements. 'We have heard much of other sources of national prosperity, but we seem to forget, that no nation can be permanently happy and powerful, that does not unite a judicious system of agriculture, to the advantages of domestic manufacturing industry, and to the benefits of foreign commerce.'[64] Many, including Arthur Young, were sceptical about the likelihood of the idea being accepted by the government, mainly because of the expense, and it was similarly rejected by Fox and his Whigs on the grounds of cost and the opportunity it offered the government of creating more jobs for its supporters. However, in spite of these objections, the bill to set up a Board of Agriculture, to cost no more than £3,000 a year to run, was passed on 15 May 1793. We do not know whether Coke was in the House that day or whether he was loyal to Fox when it came to the vote or supported the agricultural interest. Once formed, the Board's first president was Sir John Sinclair and Arthur Young the first secretary. The work of the secretary was overseen by a Board consisting of prominent representatives of church and state, as well as the president of the Royal Society and the surveyors general of woods, forests and crown lands. Below these were the 'ordinary members', made up of the 30 leading agriculturalists of the day, of whom Coke was one. Its first meeting was on 4 September 1793, but no minute books survive from before 1797. The Board met weekly during parliamentary sessions (usually November to June) although there were frequent occasions when they were adjourned for lack of a quorum, and often the attendance was below 10. A typical meeting was that of 13 February 1798, when seven members heard papers by leading agriculturalists such as the Northumberland

stock breeder Mr Culley on 'the size of cattle' and the land surveyor and writer
Nathaniel Kent on cottages. Mr Amos's paper on drilling was well received, and one
of his drills for beans was ordered.[65] Premiums were offered for such activities as
cottage building, draining, irrigation, management of cattle and sheep and
comparisons between the use of horses and oxen. Attendance at the Annual
Meetings was better, with 28 members present in April 1798. One of the duties of
the Annual Meeting was the election of ordinary members. The five members with
the lowest attendance over the previous year were automatically removed, allowing
five new members to be elected. Coke was not alone in being infrequent in his
appearance at meetings, but there were many occasions over the life of the Board
when he was one of those who lost his seat. Invariably, however, he was re-elected
the following year. While he was present at few meetings in the early years of the
Board, in 1805 he was elected a vice-president and attended three times in 1806. The
first occasion was 25 May, when communications on the cure of smut in wheat and
the use of kohlrabi as feed were read. A ballot was held to decide who should receive
the gold medal for the best essay over the previous year.

One of the major achievements of the Board of Agriculture was the publication
of a series of county reports for nearly every part of the United Kingdom. In some
counties such as Norfolk there were two. The first, by Nathaniel Kent, published in
1796 was superseded by one by Young in 1804, because, according to Young, the
introduction of a new breed of sheep and the increase in the practice of machine
drilling, rather than broadcasting or dibbling by hand, had made the previous report
redundant.[66] Certainly, Kent had not mentioned drilling, while Young devoted 14
pages to it. Kent favoured the old Norfolk breed of sheep, while Young preferred to
Southdown. A second, much expanded version of Kent's report was published in
1813 without explanation. The reports' main aim was to describe the state of
agriculture and the potential for improvement to provide for increased wartime
demands. Not surprisingly, as we have seen, the achievements of Coke at Holkham
are referred to.[67] Several county reports were referred to the Board for consideration at their meeting of 25 May 1806, and there was discussion as to the best
form in which to publish Humphrey Davey's lectures entitled 'The Chemistry of
Agriculture', which had been given to the Board, and whether a room in the Board's
premises could be converted to a laboratory for him to use in his analysis of soils.
Then came the election of honorary members, in which Coke proposed his son-in-
law Thomas Anson, but he failed to receive enough votes. Coke's second recorded
appearance was on 25 February 1806, when he supported the candidature of Mr
Beck, his tenant at Lexham, for a gold medal for his irrigation of water meadows
near his farm. In May Coke also attended, when he received the medal for irrigation,
possibly on behalf of Mr Beck, and in reply said that he hoped the Board would all
come to his sheep shearings, when they could see 30 acres that had been worth
nothing and was now greatly improved by irrigation.[68] Early in the same year,

Arthur Young as secretary wrote to him asking him to speak to Fox to ensure his support for the annual government grant of £3,000 to the Board.[69]

What were Thomas William Coke's farming achievements, and what do they reveal about the man? It is certainly true that he inherited a well-run and progressive estate where most of the open fields and many of the commons had been enclosed and the tenants were already well known as wealthy gentleman farmers ready to experiment and innovate. But it is also true that he invested heavily in the infrastructure of his estates, providing his tenants with new buildings and well laid-out fields and enclosing the last remaining commons by the end of the Napoleonic Wars. His leases continued and improved upon earlier traditions, encouraging new farming methods and requiring tenants to carry out improvements such as marling and to farm in such a way as to not exhaust the soil, by rotating cereals and fodder for stock. He encouraged the new fast-maturing sheep breeds and took a genuine interest in the quality of his stock and in cross-breeding programmes. We know that he spent many hours wearing a shepherd's smock in the sheep pens engaged in the laborious task of studying the conformation of his ewes and rams and choosing which should be put together. His Home Farm was rightly famous for the experimentation that took place on it, and Blaikie was a frequent contributor to the farming press.

However, in many ways these activities were typical of a sizeable group of landowners of the time. For instance, articles in the *Annals of Agriculture* show a number of gentleman farmers producing meticulously collected measurements to compare the merits of the different breeds of sheep. As agricultural prices rose through the Napoleonic Wars agricultural 'improvement' was seen by many to be an extremely profitable venture. What was different about Thomas William Coke was his eye for and love of spectacle. This was seen in his organisation of the festivities for the centenary of the Glorious Revolution, but it is also evident in the sheep shearings. Only the Duke of Bedford ran shearings that could in any way compare with Coke's and his did not continue for so many years. The Holkham shearings and the lavish hospitality which accompanied them gave publicity to Coke and his farms which certainly led to an exaggeration of his achievements which he made no effort to stifle. Indeed in 1817 Arthur Young felt it necessary to write to the *Bury and Norwich Gazette* correcting Coke's assertion that the agriculture on his estates was 'execrable' when he inherited and that he had been responsible for the establishment of the wool fair at Thetford. However, he wisely ended his letter by adding that 'no man holds Mr Coke in higher estimation than myself'.[70] We have also seen that this type of exaggeration was something Coke was prone to, but it does not detract from his sizeable contribution to the publicising of new agricultural methods or his enthusiasm for experiment, particularly in sheep breeding. It did much to enhance his political standing as both Whig and patriot, while confirming his reputation as farmer and improving land owner.

Chapter Six

THE ROAD TO REFORM: COKE'S LATER PARLIAMENTARY CAREER

After six years out of parliament, Coke had been preparing for the 1790 election for at least a year. One of the aims of the 1788 celebrations had undoubtedly been to enhance his political standing locally. The Prince of Wales had been prevented from attending the festivities because of his father's illness and over the following year the likelihood of a regency being declared increased. This would precipitate a general election and Coke needed to be ready for it. Lists for canvassing were drawn up. Humphrey Repton, who was to become better known as a landscape designer and who had already been employed as a political agent by William Windham in the 1780s, was commissioned by Coke to draw a map entitled 'A general view of the Influence operating in the County of Norfolk'. It is perhaps the first piece of market research produced cartographically for political ends (Colour Plate 5).[1] The map was compiled Hundred by Hundred and the number of voters in each Hundred was listed. Sometimes the number for large villages or towns was given separately. Significantly, the names of those with incomes of over £1,000 a year and who were thus deemed to influence more than 20 voters were also given, along with the candidate they were likely to support. As well as the dozen or so major country houses owners, 150 landowners were thought to exercise influence over the remaining 4,850 voters. Of particular interest is the marginal illustration, which clearly depicts the 'influences' at work in elections. The Whig cap of 'liberty' is there, as is the Whig motto 'wisdom with patriotism', on the banner, but so also are the trappings of the corrupt electoral system of the time which they claimed to abhor – bribery and unlimited alcohol. One voter is so drunk that he is being taken to the poll in a wheelbarrow.

In the event, the king's recovery delayed the dissolution of parliament until June 1790, and what might have been an expensive election for Coke was in the end averted by the withdrawal of Edward Astley from the contest. Thomas William Coke and his Tory rival, John Wodehouse, were thus duly elected, with William Windham retaining a seat for Norwich.

Coke returned to parliament at a time when the whole political map was about to change. The election had confirmed that the young William Pitt was now firmly

Plate 9: Marginal drawing from Humphrey Repton's map of political influence in Norfolk drawn before the election of 1790. The cap of 'liberty' is there, but so are the trappings of the corrupt electoral system of the time such as bribery and unlimited alcohol (one voter is so drunk that he is being taken to the poll in a wheel barrow). (Holkham MS: F/TWC 13)

in charge with a majority of over 150, and the French Revolution was about to throw into sharp relief the arguments over 'liberty' which were to divide the Whigs. Fox and his supporters saw the overthrow of a despotic monarch as the dawn of a new age, with Fox exclaiming in a letter: 'How much the greatest Event it is that ever happened in the World! And how much the best!'[2] More pragmatically it was concluded that domestic turmoil would prevent France presenting a threat to Britain. Even Prime Minister Pitt declared in parliament early in 1790 before its dissolution, that the 'present convulsions in France' would 'sooner or later terminate in general harmony and regular order'.[3] In 1791 the newly formed Norwich Revolutionary Society met to celebrate the fall of the Bastille. A new era of reform politics within the city can be dated from this event, and during the following year more societies were formed.[4] The more cautious amongst the Whigs were doubtful. Burke from the beginning warned: 'Whenever a separation is made between liberty and justice, neither is in my opinion safe.'[5] William Windham soon became one of the most vehement opponents of the Revolution, warning early in 1790 against the 'swarms of these strange impractical notions that have lately wafted over to us from the Continent, to prey like locusts on the fairest flowers of our soil, to destroy the boasted beauty and verdure of our Constitution'.[6] However, the divisions which events in France were to cause were not yet obvious, and relations between the friends remained cordial through 1790. Windham recorded in his diary dining with Coke, Fox and Burke in April of that year.[7] The House of Commons that Coke entered in November 1790 was thus not yet profoundly divided over its attitude to the Revolution, still to enter its bloodiest stage. In December 1792 Fox introduced a motion, which was defeated, to recognise the Republic of France. It was not until February 1793, with the declaration of war on Britain by France, that the splits within the Whigs became really serious, and Windham finally broke with Fox, while Coke remained obstinately loyal. In Norwich, the intelligentsia were deserting the reform clubs as events in France became more violent and instead found their support amongst those suffering from the city's economic decline.

In the autumn of 1790 these events were still several years away. The county electors were primarily interested in local issues and personalities and Coke promised that he would always uphold their interests. While parliament had been in recess agreement had been reached with Spain to drop her claim to the right to colonise along the entire Pacific coast of America as far as Alaska. While no shots had been fired, a naval mobilisation had been necessary and the stand-off had cost £3,133,000 – a sum that Pitt proposed to pay off in four years by temporary taxes. For this he introduced a supplementary budget in December which included some new duties. It was issues of local rather than national importance which determined which parts of Pitt's proposals Coke could support. On 21 December he spoke opposing the malt duties, which he claimed would have a depressing effect on the barley growers of Norfolk, but at the same time he supported a tax on dogs, which, enthusiastic sportsman that he was, would have certainly affected him personally.[8] During late 1790 and into 1791 Pitt was trying to broker an initiative across Europe which would limit the expansion of Russian power, but support from other governments was lukewarm and the threat of war with Russia loomed. Along with the other Foxite MPs, Coke was vehemently opposed to war, stating on 29 March 1791 that he did not see that Russia represented a threat and that Members of Parliament would 'not do their duty to their constituents if they were to load them with additional burdens for the maintenance of interests which were neither explained or understood by that house'.[9] Indeed, by April 1791 Fox was advocating an alliance with Russia. Although both these motions were defeated, it was clear to Pitt that he was losing support, and he was forced to accept that Russia would have to be allowed to keep its Turkish conquests. Empress Catherine was said to display a bust of Fox in St Petersberg and was heard to observe that 'dogs that bark do not always bite'.[10] In spite of this climb-down, Pitt retained his parliamentary majority and the Foxites remained an opposition, if an increasingly disunited one.

The break between Fox and Burke over support for the French Revolution came at the beginning of May 1791, and a fault line had been opened which could only widen. As events in France became more violent, Fox grew isolated but Coke continued to stand stalwartly beside his friend. Pitt, however, remained confident that Britain could remain neutral and need not be drawn into affairs across the Channel. He refused to support the calls of the rulers of Austria and Prussia for a united European effort to restore the French king. Further divisions were obvious in 1792 over a move for parliamentary reform which many thought ill-timed. Windham famously commented that 'no-one would select the hurricane season in which to begin repairing a house';[11] a sentiment which was supported by Coke.[12] Unrest in the country was increasing, and Jacobin clubs, particularly in cities such as Norwich, which was suffering from a decline in its textile manufacture, were flourishing. In the face of this upsurge in radicalism a 'loyal address in favour of a royal proclamation against sedition' was proposed in May 1792, but Coke, along with most other Whigs, refused to sign. Coke spoke in the House, saying that he had confidence, 'from what had fallen under his own observation, that the government had done all that it was in its power to bring them [the rioters] to punishment'.[13] Nevertheless, the government obtained its majority and the publication of the proclamation was followed by a flurry of loyal addresses of support. A 'very numerous and highly respectable meeting' of addressers was held in Norwich in late June which was attended by most of the county's MPs, and a proclamation in support of the constitution and against republicanism was read and supported by Charles Townshend, MP for Great Yarmouth, declaring that 'We are a free and happy people.' Robert Buxton, MP for Thetford, was more cautious. While supporting the present constitution, he thought there was room for reform. Coke, on the other hand, thought the proclamation unnecessary. Windham's address was described as 'manly, energetic and elegant', but typically for him, 'too long to quote in full'. Although he supported the proclamation he too was in favour of reform, but not of the corresponding clubs with links to France. The Proclamation was finally signed by 300 people.[14] 'Windham spoke well', wrote Coke to Parr. 'I only wish it had been in a better cause. I must confess I do not feel myself impressed with the terrors of impending evils with which he said his mind was filled.' Coke claimed to be the only one present who did not sign the address.[15]

However, the situation in France was seen by the government as increasingly dangerous, with liberty replaced by fear. Poor harvests at home fuelled discontent and encouraged recruitment to the radical societies in Britain. Rumours of French agents at work only served to make the government more nervous. Pitt recalled parliament in December 1792 and the local militia were called out in many counties. Fox and his friends were convinced that what they saw as an over-reaction was in fact an attempt by the king to impose arbitrary rule. Fox's speeches in the Commons became more extreme, as he accused the government and the king of acting together to destroy civil and political freedom. These views chose to ignore the atrocities which by now were being carried out in France in the name of liberty. In January the French king was executed and by February France and Britain were at war. This led to the final split within the Whig party. Fox and his associates, who included Coke, refused to accept the need for war. Coke wrote to Samuel Parr from London in the summer of 1793 that the country was 'about to be plunged into an unnecessary and calamitous war' and that the government was 'working on the fears of the timid'.[16] Burke and Windham on the other hand were prepared to work with Pitt. Parr, in his reply to Coke, wrote: 'It gives me the most agonising feelings to find our honourable and illustrious friend, Mr Windham, so devoted to bad men and enslaved in a bad cause.'[17] In spite of this deep rift and the desertions from his party, Fox wrote from Holkham in 1794 of the possibility of rebuilding the Whigs 'from small beginnings into a real strong party such as we once were'.[18] However, in reality Fox was tired and disillusioned. His financial affairs were in disarray and it was only through the efforts of his friends that a fund, of which Coke and two others were trustees, was set up to rescue him. From 1794 to 1801 he virtually withdrew from political life and instead indulged his scholarly interests, both in the classics and the writing of his interpretation of history.

Coke continued to believe that war could be avoided and that a return to peaceful coexistence was possible. In April 1794 during the debate on Harrison's motion to raise money for the war by taxing government placemen and pensioners, Coke made clear his disapproval of the situation, pointing out the distress it was causing in Norfolk and particularly 'the great and opulent city of Norwich', where poor rates had risen from £17,000 to £21,000 in that year. However, the idea of taxing government appointees was something which he supported. A note of cynicism crept into his contribution to the debate when he commented that he had no doubt that the minister who had proposed it 'had patriotism enough to give up all his sinecure emoluments for the public service by way of animating others to follow his example'.[19] In the same month a County Meeting was held in Norwich to open a subscription to help pay for the war, again opposed by Coke and James Mingay, an MP for Norwich.[20] Calls for peace continued, and Coke supported Wilberforce in the anti-war motion he introduced on 24 March 1795.[21] In January Norwich had petitioned against 'a war which had nearly annihilated the Manufactories and Trade of this once flourishing City and consequently reducing the majority of its inhabitants – the industrious poor – to a state of extreme distress'.[22] Later that year, in November, Coke presented a petition from Great Yarmouth against the latest repressive legislation proposed by the government to control public meetings.[23]

As the war dragged on the economic situation worsened. In March 1796, Coke spoke against the proposed closure of the banks, a measure designed to prevent a run on gold.[24] Not surprisingly, however, his main concerns were those of the agricultural interest. On 31 March 1797 he argued that farmers should be able to export their grain, 'otherwise they are unable to pay their rent'.[25] This was a year in which the price of wheat had dropped in Norwich from a record £5 a quarter in 1795 to £2 – a price not low compared with pre-war returns, but below that which farmers and landlords were coming to expect and on which landlords now based their rents.[26] Over the following months he argued for lifting the ban on the export of barley. By the middle of May he felt that there was no need for any restriction, as the prospects for harvest were looking good.[27] In fact prices were rising fast and by 1800 reached a new high of £6 10s a quarter, suggesting that there were in fact real shortages. The following year he was arguing against a new land tax. 'This calamitous war has much injured the value of land, the Norwich manufacturers are almost ruined', and his constituents, 'the landed interest in the county are very much alarmed'.[28] It is surprising that Coke chose to claim that farming and the price of land was in depression. While it was true that Norwich industries were suffering, in reality farmers and landowners were doing well out of the war, and for this reason alone many of his supporters might see an advantage in the war continuing. However, no one wanted an increase in taxes and here Coke was on far surer ground when opposing them.

There were other matters of concern during the war years. The subject of parliamentary reform continued to come up and Coke continued to vote for it, although even he was beginning to feel that the time was not right. A petition with 3,700 signatures was sent from Norwich to parliament in 1793 in favour of annual parliaments and adult male suffrage,[29] but as we shall see, Coke's ideas of reform did not include such radical change. In February 1795 he voted to pay the Prince of Wales £100,000 towards clearing his debts, although he thought the Duchy of Cornwall should have been sold first.[30] It was with more local and rural matters that Coke as an independent County Member felt most comfortable, and a year later he introduced a bill to shorten the shooting season, so that it began a fortnight later, arguing that in Norfolk, the corn was often not off the land by 1 September.[31] This may seem a surprising approach from such a keen sportsman, but no doubt farming prosperity, and thus the ability of his tenants to pay their rents, came first. The subject came up again in February 1797, when he pointed out that 'sportsmen did a great deal of damage to corn between the first and fourteenth of September'. However, this time he was outvoted by a majority of 38.[32]

In 1796 there was another election in which Coke and Wodehouse were returned unopposed, but not until there had been efforts to find a candidate to stand against Coke. His election address was highly anti-government and angered many of his would-be supporters. It was seen as dictatorial and arrogant. In it he boasted that 'I have not a single vote on my conscience for either of the two most disgraceful and ruinous wars this country has ever been engaged in.' A speedy peace and rigid economy were needed 'to return this country to its former flourishing state'.[33] His opponents declared that the county would 'not suffer Mr Coke, because his resources with respect to fortune, are so commanding, to cram his political creed down their throats without an indignant struggle on their part'.[34] Coke was spared a rival candidate mainly because Wodehouse, who was also seeking re-election, did not want the expense of a contest and so agreed with Coke not to support anyone standing against him. Various possible candidates were approached, and finally Sir Thomas Hare, the owner of the fen-edge Stow Bardolph estate but 'little known in the county', agreed to stand as 'a zealous supporter of the king and Constitution'. Finally he too withdrew, 'convinced that he had no chance of success'.[35] In the event no credible challenger could be found, and Coke and Wodehouse were duly re-elected.

The withdrawal of Fox and his close supporters from parliamentary activity was agreed in the autumn of 1797. Feeling that the independence of the House of Commons had been broken by the number of royal placemen it contained, the Foxites despondently concluded that the system of party politics was obsolete. While the split created a rift between Burke and Fox which was never healed, Windham's friendship with both Fox and Coke survived. Windham, who was now secretary for war in Pitt's government, wrote from Felbrigg in 1796: 'My connection with Mr Coke was so strong that no political Reason ... can induce me to break through it. ... I have so much respect for personal connections as opposed to political ones.'[36]

In spite of his reluctance to support the war, Coke accepted that the country needed defending against the possibility of invasion. In 1795 he predicted that if efforts to secure a peace failed, 'the consequence would be the uniting of every Englishman in the vigorous prosecution of the war'. By 1798 the danger seemed very real, and, as we have seen, Coke raised his own Norfolk cavalry, which was disbanded when peace was made in 1801 and finally ratified in the Treaty of Amiens in 1802. Although officially out of office, Pitt was involved in the negotiations, and was as committed to peace as were his political opponents. It was his former cabinet colleagues, including William Windham, who felt that too much was being sacrificed. Pitt defended his position on the House of Commons in November 1801 in the simplest of terms: 'If we had retained all our conquests, it would not have made any difference to us in point of security ... They would only give us a little more wealth; but a little more wealth would be badly purchased by a little more war.'[37] Windham's opposition to the peace made him extremely unpopular in his constituency of Norwich, which had suffered severely from the contraction of its trade during the war years. In the election of 1802, he lost his seat, having represented the city for 18 years, but instead returned to parliament as the member for the pocket borough of St Mawes in Cornwall, owned by his fellow opponent of the peace, Lord Grenville.

During these years Windham found an unlikely supporter in the form of the radical, William Cobbett. Back in England after several years in America escaping possible trial by court-martial, he found himself invited to dinner by Windham and his supporters. Knowing that he was against moves to make peace, they were anxious to have his journalistic support. Windham and a group of friends then agreed to give financial support to Cobbett to launch his, *Cobbett's Political Register*. Although the paper supported the government in its opposition to the peace, it soon began to take an independent line. In a letter to Windham of June 1803, Sir Thomas Amyott wrote: 'Cobbett's last paper [in which he had predicted national bankruptcy] gave a great deal of offence to those who were before well-disposed to his opinions: so much so that I understand that many have given or intend to give orders to decline taking his paper in the future.'[38] In fact the paper began to take an increasingly anti-government tone and Cobbett had to face a charge of seditious libel. Although Windham was prepared to be a character witness, he was becoming embarrassed by the independent line taken by his protégé. Finally, in February 1806, he broke the link, writing in his diary: 'Came away in carriage with Fox: got out at end of Downing Street and went to office, thence to Cobbett. Probably the last interview we shall have.'[39]

It is clear that in the 10 years following Coke's return to parliament he attended every year, often for several weeks in the spring, speaking occasionally and voting regularly with the Foxite Whigs. Not only did he play his part in voting on issues of national importance such as parliamentary reform and the funding of the war effort, but he also brought forward issues of local concern, such the game laws and the export of grain. He also brought the attention of the House to the economic problems of Norwich resulting from the breaking of its trading links in the course of the war. He was unquestioning in his loyal support of Fox, but accepted that once war had been declared the country needed defending.

The dissolution of parliament at the end of June 1802, following swiftly after Windham's rejection by Norwich, led to one of the few contested elections in the county. The rapidly changing political affiliations of the previous years meant that the Whig ascendancy in the county could not feel as confident as in the past. Alongside Coke was Sir Jacob Astley as the second Whig candidate, following in the footsteps of his recently deceased father, Sir Edward. Sir Jacob was an unpopular landlord, and although not a committed supporter of Fox and Coke was at least anti-Tory, which was enough for him to gain Coke's support. It was hoped that Coke's immense popularity would be enough to carry them both through. Martin Rishton, Coke's friend and political supporter, wrote in August 1803, that the yeomanry: 'in the hour of need neglected all their interests to support the object of your choice [Astley] who was to the greatest part of them obnoxious to the highest degree'.[40] Again we see the working of local political influence overriding political or national issues. In the event, both Coke and Sir Jacob were returned, beating the Tory, Colonel Wodehouse, in a contest that cost them all £35,000.[41] Nationally the 1802 election passed off quietly with few changes in the political balance in parliament.

The peace established by the Treaty of Amiens was short-lived, and by the end of May 1803 war was again declared, and with it the fear of invasion returned. A bill providing for the enrolment of volunteers in local militia was passed, but Coke refused to be involved. Despite his support of the yeomanry before the peace, he believed, along with Fox, that Napoleon was unlikely to attempt invasion. Coke's lack of concern, and of leadership in raising a militia, coupled with his continued absence from Norfolk lost him considerable support amongst his constituents. It stood in stark contrast to the enthusiasm of Windham, who, now out of office, organised local volunteers at Felbrigg and provided them with a uniform. He also inspected the Norfolk coastline in an attempt to encourage similar forces in other coastal villages.[42] Coke was finally persuaded to return to the county, and by the end of the year had armed the Wells volunteer infantry at his own expense, and had been appointed Captain of the Holkham Gentlemen of Yeomanry Cavalry. In January he was promoted to Lieutenant-Colonel of the Western Regiment of Yeomanry Cavalry, while Windham commanded the Cromer and District Battalion of Norfolk Volunteer Infantry. In a change of mood, 'a grateful public' inaugurated a public dinner in Coke's honour.[43] In April 1804 he voted in the House for both the Volunteer Consolidation Bill and the Defence of the Country Bill, showing that his earlier scepticism about the value of the local militia had been somewhat overcome. The return of the country to a wartime footing meant that more revenue needed to be raised. One proposal included the taxing of agricultural horses, to which Coke, not surprisingly, strongly objected. He pointed out in the debate of 12 March 1805, no doubt with some exaggeration, that such a tax would 'take away all inclination to agricultural improvement'.[44]

William Pitt and Charles James Fox, the two men who had dominated politics since Coke had entered parliament, both died in 1806. The death of Pitt at the end of January resulted in the formation of a government by Lord Grenville. The Grenville and Fox groups had little in common except a desire to make peace with France, to gain recognition of the rights of Roman Catholics and to abolish the slave trade. On this basis they agreed to work together. Grenville therefore, with these three aims at the forefront of their political programme, assembled what was known as the Ministry of All the Talents, with Fox as foreign secretary and Windham reinstated as secretary for war and the colonies. Now at last Fox had the opportunity to fulfil those projects in which he passionately believed. However, Mrs Stirling claims that one of Fox's first acts on entering parliament was to offer Coke a peerage, which he refused.[45] Fox was already an ill man and this, along with the knowledge that moves towards Catholic emancipation would be opposed by the king, meant that he made little effort to press this cause. Instead, peace with France and the ending of the slave trade were his first priorities. Wilberforce acknowledged the importance of Fox's commitment and contribution to the abolitionist's campaign, and the ending of the slave trade was accomplished shortly before Fox's death later in the year. Although Coke's friendship with Fox was close and he had been a very regular visitor to Holkham, Coke is not listed amongst the chief mourners at Fox's funeral. However, he certainly was one of his most loyal and uncritical supporters, who had allowed himself to be a trustee of the fund to raise money to pay his debts. The death of Fox meant that a colourful character was lost from the Commons and also that the House lost some of its appeal to Coke, who wrote at the time: 'I do not only mourn him as an individual, but as the greatest man in Europe, who might have saved this country from impending ruin and the shedding of torrents of human blood.'[46] Coke's admiration of Fox was shared by all in his political grouping. Dr Parr had described Fox in 1796 as 'the ablest statesman now living and a better patriot or more honourable man never did live'.[47] He was soon busy writing a *Life of Fox*, which he wished to dedicate to Coke. The work, finally published in 1809, was a rather curious selection of obituaries and random notes which hardly lived up to its title.[48] Busts of Fox were to be found in all the great Whig houses, including Holkham. At Felbrigg Windham displayed busts of both Fox and Pitt (both of which are still there), illustrating the difficulties he had in choosing which to support: a situation which had resulted in his nickname 'weathercock Windham'. Coke's friend and fellow enthusiast for agricultural improvement the Duke of Bedford even named his new-born son Charles James Fox Russell in the hope that one day he would enter parliament and follow in his namesake's footsteps.[49]

Even at this time, Coke's concerns were local as well as national, and he was influential in obtaining from the state the gift of Norwich Castle to the city. At the time it contained the county gaol, but later in the nineteenth century it became a county museum. He was disappointed that his Whig friends, now in power, did not pursue their earlier plan to tax private brewers, a scheme that would have benefited the barley growers of Norfolk. 'He regretted opposing measures put forward by the government's present ministers as they had his full confidence who, if the country were to be saved, were the only men who could afford such a salvation.'[50]

When the new government failed to negotiate a peace with France, parliament was dissolved in October 1806 and a general election called. Coke's fellow Whig and Member of Parliament Sir Jacob Astley decided not to contest the seat again, so Coke and Windham, reconciled by the political regroupings after the death of Pitt, agreed to stand. John Wodehouse determined to stand against them as a Tory. The cost of a contested election meant that every effort was usually made to reach an understanding between candidates by which one stood aside, but as none of them were prepared to give ground an expensive confrontation became inevitable. All three candidates published letters seeking support in the local press. Windham knew that he began at a disadvantage. His support for the war had been unpopular in the county, and his position as a member of the cabinet meant he did not have time to promote local issues. In his letter to voters of 4 October he felt it necessary to apologise for the fact that public business would prevent him coming to Norfolk to canvass as much as he would have hoped.[51] There followed one of the liveliest elections the county has ever seen. Both sides campaigned viciously. Wodehouse's supporters focused on portraying their candidate as standing for the independence of the county against Coke, whom they described as a 'perpetual dictator', a description that Coke would have rather enjoyed. Letters in the press complained of Coke's arrogance ('the electorate may be led, but they would not be driven') and of Windham's excessive use of ministerial influence.[52] Windham cannot have helped his cause when he admitted on 1 November that he might be elected for 'another place' (a government-controlled borough), if he failed to gain the Norfolk seat.[53] If Wodehouse could accuse Coke of undue electoral influence, Coke's supporters charged the Tory magistrates with threatening publicans that if they did not support Wodehouse, their licences would be withdrawn.[54]

At 36, Colonel Wodehouse was a much younger man than his adversaries. He was popular within the county, and with his wife's fortune and as heir to the Kimberley estates he was wealthy enough to fight a vigorous election campaign. Amongst Wodehouse's supporters, unusually for the period, were two ladies, Mrs Berney of Braconash and Mrs Atkyns of neighbouring Ketteringham. Both were widows of substance who dressed themselves, their carriages, coachmen and footmen in the Tory colours of pink and purple and canvassed enthusiastically around the city. A boisterous prank, typical of the sort of activity at election time and possibly encouraged by Coke's nephew Edward, involved the recruiting of two notorious Norwich ladies of the night who were dressed to impersonate Mrs Berney and Mrs Atkyns, placed in a carriage and sent out into the streets, only to be mobbed by the crowd, turned out of their coach and attacked, all adding to the excitement of the pre-election days. This was regarded as an insult by some of the Tories, particularly Mrs Berney's son, Thomas.[55]

The election finally took place between 13 and 18 November. Because of the uncertain weather and the shortness of the daylight hours, Whig supporters were requested to go straight to the Shire Hall, rather than gather for a pre-election meeting outside the city, at Mile Cross. During the morning of the first day 'country gentlemen of both parties entered the city at the head of their adherents and tenants preceded by colours, drums and music.' Thus began a

> contest for the representation of the county of Norfolk which is likely to
> be as severe, as expensive and as vehemently continued on both sides as
> any within recollection. On one side the great and merited popularity and
> personal weight of Mr Coke, the reputation of Mr Windham's talents and
> by the interest inseparably attendant upon the situation of a member of
> the administration. On the other the long-known, long tried and long
> respected services of Colonel Wodehouse's family.[56]

Although Coke was always well in the lead, the gap between Wodehouse and Windham was much smaller. Even Coke could not crack the support given to Wodehouse in his own locality. Finally, however, Windham crept ahead and Wodehouse withdrew. There is no doubt that it was Coke's popularity and influence that ensured Windham's victory. His unpopular support of the war, his distant, academic manner and the fact that his ministerial duties had kept him out of the county were in great contrast to what Bishop Bathurst described as Coke's ability to get on with all ranks of people, 'his benevolence, and the ingenuous frankness of his disposition'.[57] Coke and Windham were declared elected and were chaired round the Castle Hill and the Market Place accompanied by a hundred friends on horse-back and a procession on foot. They made speeches and gave dinners for their supporters.

However, Thomas Berney felt the insult offered to his mother and her friend by the impersonation incident deeply and would not let the matter rest. There were plenty of violations of electoral procedure with which all three candidates could be accused. With the support of five other freeholders he presented a petition enumerating charges including bribery and corruption and concluding that the poll was irregularly conducted. Parliament ordered that the petition should be considered the following February. Feelings between the two camps ran high, with the Whigs accusing the Tories of equal extravagance in the entertaining and transporting of supporters, whilst at the same time claiming that such behaviour was no different from that usually practised in Norfolk and elsewhere. Hearing the petition began on 6 February and no expense was spared by either side. Distinguished lawyers were employed and witnesses brought to London from Norfolk. It soon became clear that both sides had committed innumerable offences, and after several days the election of Coke and Windham was declared void and their seats vacant. However, both men were soon back in parliament. Windham took his seat as member for the Rotten Borough of New Romney and Coke's brother Edward, who was an MP for Derbyshire, resigned in his brother's favour and then stood as one of the new candidates for Norfolk. Sir Jacob Astley decided to return to politics and stood for the other seat. Wodehouse chose not to stand and so there was no need for another expensive election. Later in 1807 parliament was dissolved again, and Coke and Astley were returned unopposed, and the whole expensive incident forgotten.

Coke's role in the affair is unclear. Both he and Windham protested that they had nothing to do with the impersonation of two respectable widows by a couple of Norwich whores, but Coke's nephew certainly was involved and such high jinks were not unusual at election time. There is no doubt that there were many irregularities in the election process, typical of the period. Even those Foxite Whigs who were campaigning for parliamentary reform and the end of Rotten Boroughs saw nothing inconsistent in taking advantage of the system when it suited them. When sitting for his portrait by Lawrence in 1807, shortly after the end of the petition hearing, Coke told the painter that 'he well knew the management of elections and what might be done to occasion delay'.[58] The desire for a seat in the Commons overrode everything else and huge sums were spent both on the election and fighting the following case. Coke told Lawrence that the whole affair had cost all parties £70,000.[59]

The death of Fox had left Coke with less enthusiasm for parliamentary debate, and despite the great lengths he had been prepared to go to keep his seat, his attendance in 1807 and 1808 seems to have been very limited. The Ministry of All Talents, much weakened by the death of Fox, lasted only a few months and finally fell over the predicted disagreements with the king over Catholic emancipation. At the end of March 1807 a new government under the Tory Duke of Portland was formed, and Coke is recorded as voting against the Portland ministry in June 1807 and giving his support to Ponsonby as leader of the opposition later in the year. In May 1808 he came to the support of the agricultural interest in speaking against the substitution of West Indian sugar for barley in the distilleries. He argued that the best security for the farmers of Britain was allowing them a means of disposing of their surplus produce. 'Here the Honourable Gentleman entered into a detailed account of the management of barley farms.'[60] He continued to fight the farmers' cause through to June and again a few years later in March and April 1811. In 1809 he was able to return to the old Whig issues of anti-corruption and anti-monarchy, alongside his old comrades Sir Francis Burdett and William Windham in a debate on the affairs of the king's younger son, the Duke of York. 'He was determined to oppose corruption, what ever form it might assume; its defence he would leave to those who were likely to thrive by it.'[61] The following month he supported a motion to investigate further the Duke of York's affairs. It was heavily defeated. A week later he voted for an enquiry into ministerial corruption, which was similarly defeated.

In 1811 Coke had the satisfaction of seeing his sister's son, John Dutton (later the second Lord Sherborne), stand in a by-election as an MP for Gloucestershire promising, 'To support to the utmost of my power the glorious constitution; to oppose all revolutionary principles; to support reform in Parliament as to rotten boroughs; to oppose corruption in all its shapes; and to vote for the abolition of sinecures and pensions.' He must have been equally disappointed when the young man lost the election.[62] The same year saw the onset of the final illness of George III and the establishment of the regency. Relations between the Prince of Wales and the Foxite Whigs had cooled considerably and they were anxious that their opponents should not dominate the Prince's Council. Lord Albemarle wrote to Coke asking him to return to London to support him in 'the last great struggle'. He did, however, add that 'the Prince's conduct throughout this regency business has been such as to deserve your support.'[63]

Two Whig rallying causes were reform of parliament and Catholic emancipation, both of which were brought forward over the next few years. Fellow agricultural improver and regular visitor to Holkham, John Curwen of Workington, Cumbria, consulted Coke before introducing a bill for parliamentary reform. Coke voted for it, but again, it was heavily defeated. In May 1809 he voted for Brand's bill for reform, only to see it also lose. In April 1812 he voted for Catholic emancipation. After 1812, much of the excitement generated by the reformers subsided. The murder of the prime minister, Spencer Percival, in 1812 horrified moderate opinion and aroused fears of anarchism and the spectre of revolution. The last years of the Napoleonic Wars were marked by patriotic support for the throne, against which there was little that the reforming Whigs could do.

From 1814 a new issue which was to lead to Coke's unpopularity in his own county was the Corn Laws. By the end of 1813 Napoleon was in retreat and Britain, Austria and Prussia were invading France. In April of the following year Napoleon was forced to abdicate, the French monarchy restored and Napoleon exiled to the island of Elba. The British economy had to adapt to peacetime conditions. One of the first acts of the Tory government was to attempt to protect British agriculture by imposing a tax on imports, including corn, whilst allowing exports. During the war grain prices had risen to all-time highs, with the price in Norwich hitting £6 10s a quarter in 1800 and £5 15s in 1814. It was never below £2 during the war. Even this was considered a good price, allowing farmers to prosper and landlords to increase rents.[64] Although a Tory measure, the Corn Laws were something which, as a landowner and upholder of farming interests, Coke was bound to support, even though they could lead to higher prices for the majority of the population. Speaking in the debate on the bill allowing the export of corn, on 16 May 1814, he tried to pacify his critics and side-step the main issue by saying that he believed that the best way to promote the growth of corn and thus reduce its price was to promote agriculture. This could be achieved only by granting long leases and ensuring a good understanding between landlord and tenant.[65] Later in the year he presented almost the only county petition to parliament which was in favour of the Corn Laws.[66] Most of the others came from industrial areas where the inhabitants supported free trade.

Whilst the county's farmers might support the Corn Laws, the citizens of Norwich did not. High prices meant wages too might have to rise, which could increase the problem of unemployment in a city already suffering from a declining textile industry and the inability to export during the war years. There were fears too that Coke's agricultural activities and promotion of large farms were causing unemployment in the countryside. As the economic situation in the city worsened, fuelled by high wartime taxation, feelings came to a head in March 1815 when a mob attacked Coke in Norwich. The city was in uproar for a few hours on the evening of 16 March. The *Norfolk Chronicle* was swift to deny claims in the London papers that Holkham Hall had been burned. Instead it described the general meeting of the Norfolk Agricultural Society, which had begun with oxen ploughing matches on the outskirts of Norwich at which Coke and the heir to Windham's estate of Felbrigg, Admiral Luken, had been present. The event then moved to the Norwich livestock market, which at the time was held in the castle ditches, where cattle were being shown. It was here that a mob attacked Coke, Lord Albemarle and others with stones. They escaped to the Angel pub, where a mob gathered which was finally dispersed by the mayor and constables.[67] A letter written a couple of days later to Coke at Holkham from Sir James Smith in Norwich expressed his relief that Coke was safe, although there had been those in the mob threatening his life. 'You must not come here any more at present.'[68]

As the mob was attempting to attack Coke in Norwich, Napoleon, having escaped from Elba, was progressing north through France, gathering support all the way and preparing for a final assault on the allies. The British government now had to brace itself for further war and in April the House of Commons debated the resumption of hostilities. Coke supported Whitbread's amendment opposing the war, but he could muster the support of only 37 MPs and so lost heavily. Further votes followed, but support for the government remained strong, and the hoped-for peace and economic recovery were delayed.

Napoleon was finally defeated at Waterloo in June 1815 and a peace treaty signed in Vienna in November. There followed several years of economic and political unrest as the country adjusted to a peacetime economy while paying the costs incurred by 20 years of war. Unemployment was high as a result of an industrial slump and a reduction in the size of the Army and Navy. In spite of protection, agricultural prices fell, and in February 1816 Coke spoke against the renewal of income tax. Barley farmers, he said, were suffering most and wanted a fair price for grain so Coke called for the ending of the malt as well as income tax. Once income tax was 'admitted in time of peace, the country would never again be rid of it'.[69] In March he attacked the property tax as wicked and immoral, 'utterly at variance with civil liberty'.[70] Later in the month he voted against the army estimates. The maintaining of a large army in peacetime was unnecessary and one of the reasons, as Coke saw it, for the need for high taxation. Two days later he voted against the army estimates.[71] In further efforts to reduce government expendi-ture and to curb royal extravagance he voted in May against the Civil List and called for an enquiry to look into ways of retrenchment.[72] Meanwhile conditions in the country continued to deteriorate. Distress in the agricultural sector had resulted, so Coke claimed, in two of his tenants committing suicide.[73] Certainly, the tenancies of several of his farms changed hands that year.

Outside parliament, too, feelings were running high. In February 1816 a Hundred meeting in Lord Albemarle's district of Guiltcross and Shropham attacked the government and called for the repeal of the property tax. Albemarle reported to Coke that 'Ministers were bribing throughout the County to ensure continuation of the Property Tax.' This was followed by a County Meeting organised 'to have the appearance of a spontaneous outcry from the people rather than a few gentry'. Albemarle suggested taking the petition to Norwich market to get more signatures.

A similar Hundred petition had been signed on market days at Harling, Thetford and Brandon. In fact the meeting was controlled by farmers of all political persuasions and none. It was not attended by Coke or Albemarle, who claimed that if he went his language might be too strong, but 'the Court cannot understand language that is not a little strong'. Their petition was not a Whig challenge to the government, but simply called for agricultural relief. A second meeting was then arranged by the old Whig grouping, which produced a petition along more specific anti-Tory lines, stressing the extravagance of the court and its contempt for the distress of the country. 'It would be a fine triumph to check a corrupt and profligate court and to overturn a servile ministry.'[74]

As economic conditions worsened with the end of the war, demands for parliamentary reform grew, reform clubs proliferated and unrest increased. The freeholders had the right to demand that the Sheriff call a County Meeting and these were proliferating – an indication of an increasing interest in political affairs. Corn prices were being kept high by the Corn Laws for the benefit of the landed and farming interests, but causing suffering for the labouring population in town and country. For the first time economic distress was becoming politicised, fuelled by an expanding radical press and the production of political pamphlets and propaganda. The circulation of the radical *Norwich Mercury* trebled between 1814 and 1819, and the newspaper contained more political news and comment than previously.[75] The editor of the *Mercury*, R. N. Bacon, better known for his writings on agriculture than politics, wrote in his paper in 1826 that over the previous 40 years the face of politics had changed. Education and literacy were extending interest in the subject and the press was now influencing the whole of society.[76]

This pressure for reform gave Coke and his fellow Foxites a cause to support. On 13 May 1816 Coke spoke in the House complaining about the unconstitutional use of the Army.[77] In November of the same year a political meeting was held at Spa Fields in London addressed by the radical Henry 'Orator' Hunt, who appeared with an escort carrying the tricolour flag of the French Republic and a cap of liberty on a pike. At a second meeting a fortnight later a gunsmith's shop was broken into, but the riot was quelled by a force collected by the lord mayor. Memories of the French Revolution were still fresh in the minds of Lord Liverpool's reactionary Tory government, who feared mob violence and so proposed more repressive measures, all of which were opposed by the Whigs. While the protesters were ready to take to the streets in great numbers, the real threat of revolution existed chiefly in the minds of the governing classes. The cabinet decided to suspend Habeas Corpus and to pass measures against the holding of 'seditious meetings'. The bill was voted upon on 26 February 1817, and passed by a majority of 175, with Coke and 98 other Whigs voting against it.[78] Coke kept up the pressure a couple of days later by asking for a constitutional definition of 'force', and expressed the depressing opinion recorded in *Parliamentary Debates* that 'he very much feared that there existed a disposition and settled intention to produce a state of things in England nothing short of despotism.'[79]

Back in Norfolk a County Meeting was called for 5 April 1817 at which Coke spoke. He congratulated the sheriff for calling a meeting at a time when it was becoming obvious that the king was intent on overturning democracy and 'enslaving the country' by restricting freedom of speech and extinguishing the liberty of the press. Ministers wanted to deceive the people into believing that plots and conspiracies existed. The Spa Fields riots could easily have been controlled and were no excuse for the introduction of repressive measures. The solution was to get rid of the government.

It was in the second part of his speech that Coke showed himself as an old-fashioned Foxite, rather than a modern radical. He claimed to be the enemy of despotic government and a reformist, but, recalling events in Norwich, he did not support annual parliaments – 'Annual parliaments would be annual riots' – or universal suffrage – 'Universal suffrage would be universal confusion'. He had rejoiced in the French Revolution; now there was a league between ministers and a despotic monarch. Again he harped back to the past. Pitt had never used his powers to end the abominable slave trade; it was left to Fox when he came to power. Coke's speech did not go unchallenged – Wodehouse spoke supporting the government and criticising Coke for having described the restored Louis XVIII of France as a 'usurper'.[80]

The occasion sparked off a spirited exchange of pamphlets initiated by the Revd George Burges and responded to by Archdeacon Glover. Coke was castigated for his 'support' of the rioters. 'Would you join the already too numerous host of those who insidiously lull us into security that they may betray us into destruction?' Neither had he learned the lessons of the French Revolution. At Holkham 'where I am willing to believe you rule as much by affection as by power, your authority may be paramount', but elsewhere 'you command only by your virtues'. 'You can no longer dictate to the County of Norfolk what political opinions they should entertain.'[81]

Later that year Sir Jacob Astley died, which led to a by-election. In tune with the changing times, this election was likely to be more political than previous ones. The Whig candidate was Coke's long-standing friend and father-in-law of Sir Jacob's daughter, Roger Pratt of Ryston Hall, near Kings Lynn. However, he was not well known, and even Coke's wealth and influence could not get him elected against the formidable opposition provided by Edward Wodehouse. The result is indicative of two things, the lack of support for radical politics amongst the local freeholders, most of whom were farmers who feared the lifting of the Corn Laws, and the fact that Coke's overbearing use of wealth was making him unpopular in some quarters. The Whig's attempt to win over the farming community had failed.

During the following year (1818) Coke was back at Westminster, this time voting against the Royal Households Bill in March[82] and objecting to payment to the Prince

Regent towards the cost of the weddings of the royal dukes 'under the present distresses of the country'.[83] The game laws were still a subject of concern, with bills enforcing penalties which included transportation for repeat offences against poaching being passed almost annually for the 10 years after 1816. The selling of game was also illegal, those buying and selling game also liable to punishment. In March 1819 he supported a Game Law Amendment Bill which would make it lawful for tenants to sell game shot on their land and thus, it was hoped, drive the poacher out of the market. 'A man had an inherent right to that which was on his property', however large or small that property was. The bill, however, was defeated.[84] The selling of game through licensed outlets was finally made legal in 1823.

In February 1819 he presented a petition from the farmers near Kings Lynn pointing out the degree of agricultural distress. While, in contrast to the rest of his party, he had in the past always supported the Corn Laws, he did not agree that the price of grain should receive further support. Like many in the government, Coke seems to have regarded the Corn Laws as a temporary measure to help the transition from a wartime to a peacetime economy. Instead he proposed the lowering of taxes, in particular that on agricultural horses.[85]

Unrest continued, and meetings of those seeking reform increased. In August a meeting of between 50,000 and 60,000 people at St Peters Fields in Manchester was broken up by force, causing the deaths of 11 of the protestors and leaving about 400 wounded. In the outcry that followed it was unclear whether the meeting had been illegal, and to clarify the situation the government decided to put forward a further Seditious Meetings Prevention Bill, which came before the House on 2 December and against which Coke spoke. 'He thought the demand of ministers for those extra powers quite unwarrantable, especially, satisfied as he was, that the meeting at Manchester would have gone off as quietly as those at Norwich, Hunslett Moor [Yorkshire] and as many others of the same description if it had not been interfered with by the officious agents of authority.'[86] Coke returned to the subject later in the debate, when, amidst cheers, he went so far as accusing the government of being 'most strongly implicated in the events in Manchester'. Certainly the government was known to employ spies and *agents provocateurs* in its attempts to root out trouble. Because the government refused to hold an enquiry he was convinced that they had incited violence so that they could get their repressive legislation through. At this point the Speaker intervened to prevent further accusations.[87] The Six Acts, passed by the end of the year, regulated the holding of public meetings, encouraged magistrates to search for arms, increased the stamp duty on newspapers and tightened the laws against seditious publications, all of which were vociferously opposed by the Whigs. In the event the legislation was enforced with some restraint and its powers were rarely used.[88]

Coke continued to argue for retrenchment and opposed new taxes. His frustration with the inability of the opposition to influence policy is shown when on

18 June he complained that the House was 'corrupt' and the government, through its placemen, could always find a majority. He argued for triennial parliaments, which he hoped would reduce this influence. As always, Coke's main interest was with local matters, and, in the last month of 1819, he believed that political influence was being felt in the choice of magistrates for the county of Norfolk. The Tory Lord Lieutenant was only appointing Tories to the bench. Coke intervened and took the matter up with the Lord Chancellor, whilst at the same time publicising the abuse in the House. The matter, however, was dismissed by the House.[89]

Coke's dominating concern, and that of his electors, was the problem of agricultural distress, which he blamed squarely on the level of taxation. The depression reached its most serious level in the years 1820 and 1821, when bumper harvests resulted in a plentiful supply of grain and hence low prices, which the government blamed on overproduction. These were the only two years during Coke's ownership of the Holkham estate that rents actually declined. In February 1822, he accused the House of being 'profligate and corrupt'. To cries of 'Order!' from the Speaker Coke was forced to apologise, saying that 'he was warm and it was natural that he should be so'.[90]

It seems that at the time Coke was suffering from ill health. In 1819 his friend Dr Parr urged him to 'take care of your health during your attendance in Parliament … I must exhort you to remember, as I am compelled to do, that your advanced time of life requires caution. Few of your public duties are more delightful to me, or more creditable to yourself, than your unwearied diligence and attention to your parliamentary duties.'[91] Conditions in the country remained depressed and un-settled, and it could well be that at the age of 65 Coke, disheartened by the many defeats of the previous sessions, and missing the excitement that Fox had brought to politics, chose to take Dr Parr's advice and to spend more time on his estates.

The records of a correspondence carried on in 1820–1 between Coke and George Eyres, the chairman of the Lynn Agricultural Association (a branch of the Central Agricultural Association, formed in 1819 to combat merchant pressure to lower prices), illustrate Coke's changing attitude towards protection. The Agricultural Association was more concerned with the price of wool than that of corn and was supported by the producers of Merino wool, many of whom were to be found amongst the graziers of west Norfolk, who were losing out, as the manufacturers preferred fine German wool. Late in 1819 a duty was put on foreign imports of wool, which enraged the Yorkshire weavers.[92] In Norfolk the Association also turned its attention to the protection of cereals. Eyres wrote to Coke first in June 1820 requesting his support of the petition for further protection for agriculture which his Association would be putting to the House. Coke replied saying that he did not believe farming was any worse off than the commercial sector. This being the case the general malaise could only be blamed on the bad administration of the government. A reduction in army numbers would do more for the agricultural interest than further protection. Eyres responded by saying that notwithstanding Coke's reservations, they would continue to petition for further protection. The Association's Tory members refused to support Coke's anti-government line, and Eyres went so far as to write that 'with such an enormous debt, we could not expect any effectual [tax] relief at present.' He criticised directly Coke's anti-protection stand, which was injuring an agricultural portion of his constituents, 'for it can no longer be a matter of surprise that those members of the House of Commons who are not very conversant with rural affairs, should stigmatize us as "wicked and insane" when they see Mr Coke, the professed friend of Agriculture in opposition to his petitioners.' In his reply Coke reiterated his previous belief that it was government expenditure and thus taxation which needed to be curbed. He pointed out that the price of wool had continued to drop in spite of the latest measures to protect it. He also pointed out the lack of logic in Liverpool's belief that the low prices for corn were the result of overproduction, when one of the reasons he had put forward for introducing the Corn Laws was to boost production to prevent famine.[93] The debate rumbled on, with Coke making sure he was present in the Commons, even if he did not speak when the issue came up. At the beginning of March 1827, Frederick Madden called at Fenton's Hotel in London hoping to discuss the catalogue he was preparing of the Holkham library, only to find that Coke was not up. He had been at the House of Commons until five a.m. Later in the day Madden found him at Brooke's, where Coke told him that he would be back in Norfolk by Saturday. 'He was only waiting for the "Corn question".'[94] In 1828 a new Corn Bill was passed which reduced the level at which foreign corn could be imported from £4 to £3 13s a quarter. This represented a modest retreat from the high levels of agricultural protection given in 1815, and for a time settled the bitter disputes between the farming interests and the free-traders.

Gradually there was shift in emphasis in the petitions from the county that Coke was presenting, no longer seeking government support for agricultural distress, but promoting the belief that parliamentary reform was the only possible remedy. A County Meeting in January 1822 demanded the lowering of the duty on malt and the reduction of taxes, rents and tithes alongside greater government economy. Albemarle went as far as to propose that there should be no taxes, while Coke affirmed that 'Taxation without representation ... had been the cause of all our calamities.' Lord Suffield along with Samuel Clarke, a friend of William Cobbett, piloted the meeting towards discussing parliamentary reform, maintaining that 'without reform no relief could be obtained' and a resolution for reform was agreed.[95] In April 1822 Coke attended a meeting in Fakenham at which a hundred of the principal owners and occupiers in Gallow Hundred supported the presentation of a petition to parliament on the subject of the present agricultural distress 'and the total inadequacy of the present measures adopted for relief'. Through mismanagement, the government was driving the country to misery and ruin. If the country had been fairly represented, the salt tax would have already been repealed: instead it was the government placemen who kept it. After the meeting, ropes were attached to Coke's carriage by some of the townsmen and it was drawn out of the town 'amidst loud acclamation and the ringing of bells'.[96] Previously there had been a similar meeting in Walsingham, and in spite of opposition from the magistrates, who feared trouble, further meetings were held in many Norfolk Hundreds, always with the same aim. In April and June 1822 Coke presented petitions to parliament from Earsham and South Greenhoe, both claiming great distress in their areas and calling for retrenchment in every government department, the reduction of sinecures and all useless offices and the reform of parliament.[97] The Greenhoe petition was rejected by the Lord Chancellor, who objected to the statement that 'a standing army is kept for keeping down the constitutional spirit of the people.'[98] A County Meeting in May called openly for parliamentary reform as the only way of reducing taxation. Wodehouse was now prepared to accept that some reform was inevitable, but like other Tories, he was determined to save as much of the old electoral and political structure as possible. The demands of the landowner and farmer Thomas Beevor of Hethel and Mr Southwell, a Norwich radical, for universal suffrage were too radical even for Coke and Lord Suffield.[99]

The suicide of the foreign secretary, Lord Castlereagh, in 1822 and his replacement by George Canning marked a change of the direction of the Tory party. Unlike his predecessor, Canning favoured Catholic emancipation, and, whilst not supporting calls for parliamentary reform, he believed that this could only be avoided if parliament removed some of the abuses which were the cause of much of the discontent. As Catholic emancipation moved up the political agenda, Coke presented petitions in its favour, firstly in April 1823 from 55 Norfolk clergy and then again two years later, this time signed by 70 clergymen. He praised the Bishop of Norwich, Dr Bathurst, for his continued support, almost alone amongst the bishops, for the campaign.[100] In March 1829 a further petition was again signed by 70 clergy, including two archdeacons. The withdrawal of Lord Liverpool from public life after a stroke in 1827, followed by that of his successor as prime minister, Canning, later in the year, had led to a split in the Tories. In the confusion, the king turned to the Duke of Wellington, who could be trusted to oppose the clamour for Catholic emancipation. However, even Wellington could not stand against the pressures from Ireland, and finally, in 1829, the bill allowing Catholics full religious freedom and the right to sit in parliament was finally passed, much to the relief of Coke and his friend Dr Bathurst; it was a measure Coke claimed to have supported for 25 years.[101]

The agricultural depression continued to bite during 1822 and 1823, with the number of tenant farmers handing in their notice increasing. At Holkham as elsewhere, rent rebates of up to 30 per cent were agreed. Distress meant that parliamentary reform remained a major issue. On 4 January 1823 a County Meeting was called by the sheriff at the request of 540 'occupiers of land, tradesmen and others' to consider ways of alleviating agricultural distress. It was attended by a large majority of the county's freeholders, as well as about 500 from the various radical reform clubs in the city, and held in St Andrews Hall in the heart of Norwich. A stage was erected at one end where the country gentleman began to gather soon after the doors were opened. 'Soon after twelve the sheriff, under-sheriff and County Members [Coke and Wodehouse] arrived. Mr Coke's appearance was greeted with loud and repeated cheers.' All agreed that taxation was the major reason for distress, all needless sinecures and pensions should be abolished and beer and malt taxes should be repealed. In his speech, Coke stressed that the first step towards agricultural recovery should be the reduction in the Army, which would immediately be a great economy allowing for the reduction in taxes. So far this was traditional Whig territory, but at this point, much to Coke's annoyance, the meeting was taken over by the more radical William Cobbett, demanding further reform and universal suffrage. The meeting was a 'respectable one', but totally dominated by Cobbett, who rewrote the proposed petition in his own more radical terms. In a letter to Dr Parr following the meeting, Coke wrote that he would make it clear when he presented the petition that it did not represent the views of 'the enlightened Yeomanry or that of the people'. He went on to complain of the interference of Cobbett, describing him as causing 'discontent and ill humour. A greater enemy of reform never existed.'[102] The local papers were critical of Coke and the other gentry present for allowing the meeting to be taken over in this way. The proceedings were very confused, with at least four speakers talking at once, and people scarcely knew what they were voting for. Cobbett's petition blamed the present troubles on the 'predominance of certain particular families who ... appropriated to themselves a large part of the property and revenue of the whole nation'.[103] A reform of the House of Commons was needed to enable the adoption of certain measures. Church property and the crown lands should be appropriated to help pay off the national debt, the standing army should be reduced and sinecures should be abolished. Short-term measures to help alleviate the general distress should include the suspension of all distraint for debt and efforts to collect tithes.[104] After the meeting, 'a crowd attended Cobbett to the Bowling Green Inn from the window of which he congratulated them on the success of the day's proceedings.'[105]

In April 1823 there were several petitions from Norfolk. Although Coke supported the end of Rotten Boroughs through which government could control the membership of the Commons, he did not agree with universal suffrage, believing that power should remain with the owners of property. Unlike William Cobbett, the great Whig magnates were no more in touch with 'the common man' than the Tories. Their main aims were the preservation of the constitution, the maintenance of the union, the defence of social hierarchy and the protection of property, which meant that Coke could not support the petition movement initiated by Cobbett.[106]

When he presented the Norfolk petition to parliament he made it clear that he did not think it truly represented the views of the 'county', by which he meant the landed interest. It had been followed by several Hundred meetings producing revised versions which he also presented. In his own Hundred he had chaired the meeting, and the resulting petition bemoaned 'the misery the landed interest now laboured under which arose entirely from the actions of a servile and corrupt government'.[107]

An improvement in economic conditions in the mid-1820s led to a decrease in agitation, but depression set in again in 1829, followed by the usual round of meetings. Unrest in the country was fuelled by poor harvests and economic crisis, and, as we have seen, the Holkham estate was not immune to mob violence. The preamble to an Act of 1829 consolidating previous game laws stated that 'The practice of going out by Night for the Purpose of destroying Game was much increased in recent years', and if this was really so, it was partly the result of increasing hardship in the countryside.[108] An increase in rural violence occurred in 1830, with much destruction of machinery. Whilst determined to enforce the law, magistrates also urged farmers not to use threshing machines and to agree a local wage. While economic conditions lay behind the unrest in the countryside, that in the towns had a more political motive. Wellington's government was by now much weakened, and the death of George IV in 1830, succeeded by his more liberal minded younger brother, William IV, necessitated an election. The peaceful revolution in France and the replacement of Charles X by a constitutional monarch that year encouraged the reformers in England, and the clamour for parliamentary reform increased. The election was held in October. Having accepted the need for moderate parliamentary reform and for Catholic emancipation, both issues which had split the party, Wodehouse had lost Tory support. More importantly for local support, the Tory government had failed to end the agricultural depression. It was this that meant Wodehouse had lost the confidence of the farming voters, who then set about finding a candidate who would support their demands. They chose Sir William Ffolkes of Hillington Hall. When Coke agreed to support him, Wodehouse withdrew. While this can be seen as a victory for the voters in that they chose their own candidate, that victory still depended heavily on the establishment and the time-honoured forces of interest.[109] Coke and Sir William Ffolkes were duly elected. Their election dinner was held in the new Corn Exchange in Norwich, which was draped with banners declaring that both were 'advocates of economy and retrenchment' and would be independent members, open to 'the power of popular opinion'. In his election speech Coke drew attention to his past friendship with Fox, contrasting it with his low opinion of the prime minister, the Duke of Wellington, whose only praiseworthy action had been to see through Catholic emancipation. Other than this, he did 'not believe that he is a statesman or has one particle of freedom in his composition'. There were toasts to Lafayette, 'and may the glorious revolution [in France] of 1830 herald the grand era of European liberty (much cheering)'.[110]

The long years of Tory government, when, as Coke complained, they were able to control the voting patterns in parliament through their 'placemen', must have been very disheartening for the government's critics, and so it is not surprising that Coke's contribution to debates and mention in voting lists through much of the 1820s was so limited. In 1822 he had remarried, and as his new family grew his interest in domestic life overtook his desire to be in the Commons. It was not until Coke's friend Earl Grey became prime minister in 1831 that Coke renewed his interest in politics. A reform bill was introduced that year, and Coke expressed his delight, commenting that he had 'never indulged the hope that, at one sweep, the whole of the rotten boroughs would be carried away and the influence of the borough mongers extinguished for ever';[111] yet this is precisely what the bill proposed. Introduced to the Commons on 1 March 1831, the bill would disenfranchise 60 boroughs with fewer than 2,000 inhabitants. It removed one member from 47 boroughs with a population of between 2,000 and 4,000 and cut down the representation of combined boroughs. At a stroke, 168 seats would disappear and in their place seven large towns and four districts in London were to be given two members each. Twenty towns would have one member and 26 of the more populous counties had their representation doubled. However, power was still kept in the hands of the landed magnates. The franchise was extended to include the better-off leaseholders as well the forty-shilling freeholders in the counties. On 23 March the bill was passed at its second reading by the largest House in living memory, but in April the Tories defeated the bill at committee stage and parliament was dissolved.

After a fiercely fought election amid heightened excitement and a certain amount of rioting, the Whigs were returned with a larger majority. A second bill was brought forward and passed by the Commons, only to be rejected by the Lords. There followed riots, and the language used by the extreme radicals came very near to threats of revolution. Petitions flooded into Westminster, that from Norwich bearing 11,352 signatures.[112] This was followed by the resignations of the Whigs and an attempt by Wellington, against a background of unrest and threats of civil disobedience, to form a government. When this failed the Whigs agreed to return to office, but only if the king promised, if necessary, to create enough new peers to ensure the passage of a bill. With a few concessions a third reform bill was finally passed on 4 June 1832. Coke entered the debates on only one occasion. On 1 June, only a few days before the bill was passed, he interjected during a debate on the divisive subject of the Corn Laws to say that this was not the right time to be discussing them.[113]

The Reform Act signalled the moment for Coke to retire from politics. At 78, and having spent the best part of 55 years as a County Member, he was ready to retreat from public life. Reporting on the dinner held in St Andrews Hall to mark this event the *Norwich Mercury* commented: 'The retirement of a man like Mr Coke from public life bears the same analogy to his county that the termination of the reign of a patriotic king to the nation whose rights he has protected, whose interest he has encouraged, whose prosperity he has reared and established.'[114] Whilst this is obviously an exaggeration and Coke was by no means universally popular, particularly in Norwich as a result of his attitude to the Corn Laws, it is true that the very length of his tenure of a county seat meant that most Norfolk men and women would not be able to remember a time when he had not been one of their MPs. The dinner was attended by local and national Whig figures, including the king's brother, the Duke of Sussex, who led the tributes.

Shortly after the Reform Act became law, parliament was dissolved and an election under the new arrangements was called for the end of the year. Instead of the county of Norfolk sending two MPs, it was now divided into East and West divisions, each with two representatives. The Rotten Borough of Castle Rising had lost the right to its own Members of Parliament. The total number of MPs for the county therefore remained at 12. Hustings were no longer held in Norwich, but in a town in each division. On 15 December Coke led the procession of leading reformers, large landowners and a cortège of tenantry into the prosperous market town of Swaffham to nominate two members of long-standing Whig families, Sir Jacob Astley and Sir William Ffolkes, as candidates for West Norfolk. There beside the elegant domed butter market in the town's large market-place Coke gave his last political speech, which was designed to be a rousing one to rally his audience. His description of the Norfolk yeomanry, of whom his audience mainly consisted, as an 'influential and important body of men – persons unequalled in any country in the world and pre-eminently distinguished in the county of Norfolk for their skill in agriculture, for their ardent love of liberty, for their independence and not less for their attachment to the constitution of their country' was bound to raise a cheer. He then thanked those who had given him their support for over 60 years and explained his political priorities over that time. 'I have endeavoured to uphold the liberty and true interests of England, the freedom of mankind in general and the constitution of my country.' He had accepted no favours and had served independently. This again drew cheers from his audience. Although he had many criticisms of the long years of Tory rule that had followed the death of Fox, the longed-for Reform Act had now been passed, not by Earl Grey alone, not even by the House of Commons, but by the people, despite the opposition of the House of Lords, 'that low and wretched oligarchy who were placed in the House of Lords by that statesman' (Pitt). Now that the Bill was passed 'the cause is now your own – of no one but yourselves can you now complain.' This was followed by a vote of thanks from Mr Brampton Gurdon of Cranworth which was accompanied by 'enthusiastic shouts'. The party then retired to dine at the crown. The assembled supporters numbered over 100 and included not only the local Whig dignitaries but several of the more well-to-do Holkham tenants.[115] In 1832 the Whigs were triumphant nationally, with a majority of 370. Norfolk returned seven Whigs and five Tories. However, gradually their support dwindled and by Coke's death the conservatives were back in power, with Norfolk returning three Whigs and nine Tories.[116]

Coke chose to commemorate the achievement of limited reform in style by commissioning the famous sculptor, and frequent visitor to Holkham, Sir Francis Chantrey to create a bas-relief depicting Earl Grey and those other Whig politicians primarily responsible for its passage (amongst whom Coke placed himself beside his son) as medieval knights, in a scene reminiscent of the signing of Magna Charta. The reformers shown as knights include many of Coke's close friends such as the Duke of Sussex, Lords Spencer, Melbourne, Dacre, Lyndoch, Brougham and Denham, and Earl Grey's son-in-law, Mr Ellise, most of them regular visitors to Holkham. William IV is depicted as King John. Both events could after all be seen as of equal significance to the British constitution. Such an allegorical theme was an unusual commission for Chantrey to undertake as he claimed to hate allegory: 'it is a clumsy way of telling a story'.[117] The plaque now hangs on the wall in the great hall at Holkham. Whether it was ever paid for is unclear. It may be that Chantrey undertook it as an indication of his support for reform. Writing to Coke's friend Archdeacon Glover in 1841, apparently referring to a recent conversation, he wrote that he was 'glad to find that your recollection is so perfectly in accordance with my own'. He went on to say that he would write to Coke's brother-in-law, Thomas Keppel, 'that he may communicate its contents to Lord Leicester at once, and relieve his lordship from the necessity of keeping this money transaction uppermost in his mind'.[118]

The coronation of William IV was the occasion for the creation of new peers, and Lord Grey as prime minister was in a position to put forward names. By August 1831, the Reform Act seemed inevitable, and as Coke had made it clear that as soon as this was achieved he would retire from the House of Commons, it seemed appropriate that Coke should be offered a peerage. In August 1831, therefore, Lord Grey wrote to Coke asking for his agreement to the putting forward of his name. Coke replied, repeating the story that the offer had previously been made by Lord North in 1788 and the Duke of Portland in 1806, but that on both occasions he had preferred to stay with Fox in the House of Commons. Now, however, things were different as he had a son to succeed him, and he would be honoured to accept. But this was not to be. Creevey records that Coke made a speech in Kings Lynn at which he had pointed to a portrait of George III and described the king as "that wretch covered with blood" (meaning of course the American and French wars) an insufferable speech, particularly of a dead man'.[119] King William was not prepared to ennoble a man who had insulted his father in this way.

Coke's long period in the House of Commons saw the nature of politics change radically. There is no doubt that it was in his early years that he found public life most rewarding. The death of Fox marked for him, as for many of his fellow Whigs, the end of an era to which they long looked back nostalgically. Not only had Fox been a lively and charismatic member of the House, but after his death in 1806 his grouping was without an effective leader, and power passed out of his followers' hands. Indeed throughout the long period of opposition, until Earl Grey finally overthrew Wellington in 1830, the old stagers remained faithful to Fox's memory. Coke even mentioned him in his farewell speech, some 24 years after his death. The years after 1806 saw Coke make far fewer appearances at Westminster as, like so many others in his party, he despaired of ever again having any influence on the affairs of state. Fox's causes had inspired Coke: patriotism and the cause of liberty united in support of the American War of Independence and the initial stages of the French Revolution. Fox stuck stubbornly to the belief that war with France could have been avoided. Through the rose-tinted spectacles of old age Coke looked back to those early days in his speech in Swaffham market-place. 'When I first entered parliament this was an untaxed country, there were no poor. Every man could brew his own beer and bake his own bread. Why no more? Mr Fox foresaw the dreadful state to which this country would be reduced.' All this was the result of Tory policies. How different would things have been 'had the principles of Mr Fox been followed. We have been eaten up by borough mongers and an oligarchy living upon and out of the labour of the industrious people of this country.'[120]

Coke rarely spoke in the House, but when he did he never missed an opportunity to blame the current ministers for their handling of foreign affairs, which had led to hostilities. The themes to which he constantly returned were burdensome taxes, which were in themselves the result of wartime necessity, the corrupt electoral system by which the government ensured a majority in parliament and the many sinecures which allowed the government to reward its supporters. His dogged return to these issues at every opportunity contributed to the wearing down of the conservative element in parliament and the Reform Act.

It was also the role of an MP, then as now, to support the concerns of his constituents, and Coke often drew the attention of the House to the problems of agricultural distress, frequently speaking against the malt and other taxes, the effects of which were particularly burdensome to cereal farmers in East Anglia. He regularly presented petitions on the subject of agricultural distress from his constituents. His initial support for the Tories' Corn Laws made him unpopular in Norwich, while his lack of enthusiasm for them after 1820 lost him support amongst the farmers and their landlords.

Coke's parliamentary career and the role he cultivated for himself on his estates show him in very different lights. At home he was a confident landlord, his leading position as the owner of the largest estate in the county ensuring his social standing. His house was the acknowledged social centre of Norfolk's Whigs. Together, these ensured that his support would be sought by the leading politicians of the time, but it was as a supporter rather than a leader that his political career must be seen, and in this position he was far less sure of himself. Not a great speaker, he stuck to the few themes he understood. Signs of the arrogance with which he was sometimes accused by his Norfolk enemies led him occasionally to lose his temper in the House, earning him reprimands from the Speaker.

Whilst Coke's role remained that of a backbencher, he could be relied on to support the great Whig causes with his not inconsiderable weight and influence. Yet he was very much a Georgian rather than a Victorian politician. He saw the limited parliamentary reform enacted as a result of the 1832 bill as the culmination of Whig efforts. There was no more to be done. The propertied classes were still in control of the political system and the electorate had only been increased from 3.2 per cent to 4.7 per cent of the population. He could not agree with those radicals such as William Cobbett who saw universal suffrage as the final goal. In Coke's eyes, power should remain with the landed interest. The time of his resignation from parliament coincided with the beginning of serious questioning of this cornerstone of the old Whig philosophy. It would be up to the next generation of reformers to carry on where the Whig gentry, of whom Coke was a typical example, had left off.

Chapter Seven

'OPEN HOUSE': HOLKHAM AND ITS VISITORS

A ll were agreed that one of the most important distinguishing characteristics of a patriot was his generous hospitality and the open house he kept for his friends. Here he could display his wealth and cement the friendships which were essential to the exercise of influence both at county and national level. In this Coke excelled himself. It exactly suited his extrovert and exuberant personality. In the absence of personal diaries and papers, the descriptions and letters left by his many well-known visitors contribute to our understanding of the man. All who visited were struck by the welcome they received and by Coke's magnetic personality. When Elizabeth Coke's future husband, John Stanhope, first met his future father-in-law in 1822, he wrote of Coke, 'Besides being a very handsome man, he has an unrivalled charm of manner, combined with a simplicity of mind and character which are infinitely attractive. To me he was kindness itself.'[1]

The wife of a landed gentleman not only had to be a good household manager, but also a good hostess and Jane Coke filled this role admirably. She was reputed to be a fine horse woman who could accompany her husband's guests on expeditions around the park and nearby farms. When the famous breeder of Southdown sheep, George Ellman, and Thomas Boys arrived on a visit in 1792, they found Coke out riding in the park with his wife and daughters.[2] After Jane's death in 1800, many of the domestic duties fell to Coke's eldest daughter, another Jane, who was widowed in the same year, at the age of 23. After her second marriage in 1806 she spent less time at her father's home, and as soon as Elizabeth was old enough she took on the running of the household – a task which she found onerous and was glad to escape when the second Mrs Coke took over in 1822. In a letter from Elizabeth to John Stanhope, written in 1822, she wrote of her new step-mother, 'She probably finds what I have found for several years; that this house entails a perpetual sacrifice of all one's feelings and inclinations.'[3] Much of this was created by the large house parties that were entertained.

There were not only the house guests to be catered for, but also those who called for a tour of the house, a pastime that was already fashionable by the middle of the eighteenth century. Visitors began arriving well before the house was finished, and the wine books show that refreshments were regularly served. While most described the house as 'magnificent' others were more critical. Lady Beauchamp in 1764 wrote that looking across the marble hall from the steps into the saloon gave the impression of seeing a cold bath![4] After the death of the builder of Holkham, Lady Leicester had continued to open the house on Tuesdays, with Thursdays being her 'public' day when she personally showed people round. Wenman Coke kept up the tradition during his brief ownership of the house. The *Norfolk Tour* reprint of 1775 stated that Holkham 'can be seen any day of the week except Sunday by noblemen and foreigners, but on Tuesdays only by other people'.[5] When the La Rochefoucauld brothers visited England in 1784, they were introduced to Coke by Arthur Young at a dinner in Kings Lynn. 'Ordinarily one is able to see this house only on a Tuesday, but Mr Coke most obligingly invited us to come and spend a few days with him to see in our own time his house and estates.' The wonders of the house are fully described in their journal. Their stay must also have included tours of farms, as the journal includes descriptions of Coke's farming methods. However, rather frustratingly, the journal notes, 'This record of our Norfolk tour never got finished because at Holkham we were too busy enjoying ourselves to have time to write.'[6]

The first guide book was produced shortly after Lady Margaret's death, in 1776. This 16-page booklet listed the paintings and statues, but contained very little in the way of description. It was superseded in 1817 by a much larger production written by John Dawson of the nearby village of Burnham and containing 'a faithful and comprehensive description of every object deserving the notice of strangers'. It ran to 138 pages and included at the end a description of the sheep shearings. The tour began in the vestibule under the south portico, where there was a visitors' book to be signed and from where they would be conducted round the house by a member of the household staff, who would hope to make good money from the tips. The gardens could also be seen on application to the gardener and the family plate on application to the plate burnisher.[7]

As well as the casual visitors, there were more serious antiquarians and art critics. The Yarmouth banker Dawson Turner, who himself amassed a large collection of paintings and books, came to 'educate his taste' at Holkham in what was regarded as one of the greatest private art collections in Britain. Here, far away from the noise and bustle of his house beside the road and harbour on the South Quay, Turner could quietly study the qualities of the pictures in Coke's collection. He listened to the comments of other visiting artists, scholars and connoisseurs and noted them down in his copy of the 1817 guide book.[8] Turner also spent weeks at a time at Holkham in the library and enjoyed introducing others to it. He wrote in 1812 to the book collector Thomas Dibden inviting him to join him there: 'Mr Coke would, I know, be delighted to see you, and I could safely promise you a very rich treat in his library, which is almost wholly unexplored and contains such a treasure of early editions as I never saw in any other.'[9]

The busiest times of year for entertaining house guests were during the shooting season from November to February and the sheep shearings in late June, when beds frequently had to be hired from Wells to sleep all the guests. The La Rochefoucauld brothers noted in their journal that there were 150 guest beds (*lits de maître à donner*) and a further 150, presumably for both domestic and visiting servants.[10] While little correspondence survives from the early years, the game books show the number of Coke's Whig friends and relatives who shot with him when he was a young man, and the numbers entertained continued to be large to the end of Coke's life. Visitors for the shooting in 1822 included Captain Spencer and Lord Althorp, Sir James Smith, Lady Hervey and her sister, Elizabeth Caton, Viscount Titchfield and Coke's nephew, William Coke. In November George III's younger son, the Duke of Sussex, and his nephew, the Duke of Gloucester, joined the group, bringing the number up to 26.[11] In 1841, the year before Coke's death, when he was 87, the Revd Alexander Napier described the house as crowded for the shooting. Lord John Russell and Lord Spencer as well as the Duke of Sussex were still frequent visitors.[12]

Coke was always proud of the number of visitors he received from the newly independent United States: so much so that he collected their letters, which were later bound together to preserve them.[13] He claimed to have drunk George Washington's health every evening during the American War and to have described him as 'the greatest man on earth'.[14] Washington himself took a great interest in agriculture and corresponded both with Sir John Sinclair and Arthur Young. In a letter to the Marquis de Lafayette Washington expressed his respect for Coke and 'a great desire to know his agricultural establishment', and Lafayette flattered Coke immensely by reporting Washington's admiration to him.[15] Mrs Stirling claimed that Coke was the first Englishman to open his doors to the first ambassador sent by the United States to the Court of St James.[16] A later ambassador, William Rush, was a frequent visitor to Holkham and the sheep shearings and regularly received gifts of game from Coke. He sent letters of introduction for American visitors, who often hoped to arrive in time for the event. Indeed, there was hardly a shearing when an American was not present and they always received a particularly warm welcome, 'for are we not the same family? Do we not speak the same language?'[17] Christopher Hughes, another American diplomat, who served in Sweden and the Netherlands as well as Britain, wrote to say that the king of Sweden was an admirer of Coke. George Logan, a senator from Pennsylvania who had studied in Edinburgh, returned to America intending to devote himself to scientific agriculture. He was a member of the American Philosophical Society and corresponded with Coke on the subject of farming and the rotation of crops, and published pamphlets on various agricultural topics. However, in 1810 he was drawn into the argument over whether America should declare war on Britain in support of France and felt strongly enough on the subject to set out to Britain without any authority from the senate to attempt a reconciliation between the two counties: a mission to which Coke would have given his full support. However, he failed and was recalled to America in 1812 before he could visit Holkham.

Rufus King, another of Coke's American correspondents, also wrote expressing his view that it would be madness for Britain and America to quarrel. Although commerce between the two countries was suffering because of the war with France, with whom America was allied, agriculture was gradually improving and he thanked Coke for four cows he had been sent. His animals survived the long sea journey, and in 1821 Charles King wrote to say that some cows he had received were doing well and had won prizes at local agricultural fairs. Theodore Lyman, who was an active member of the Massachusetts Horticultural Society and twice Mayor of Boston, attended the sheep shearings in 1819 and described them as 'one of the most interesting and useful spectacles in Europe'. Similar agricultural shows were now being started in America. Senator Thomas Harper described Coke as 'the earliest and most constant friend of America' and wrote of his 'grateful recollections of Holkham' and the 'elegant and cordial hospitality' he had received. Those who came were all impressed by the lifestyle of this 'old English country gentleman'. The great banquets which were part of the sheep shearings reminded Richard Rush of the old baronial days and feudalism in 'alliance with modern freedom and refinements ... perhaps more of the romance of English History is apt to linger about an American than an Englishman'.[18] Their republican prejudices were mellowed by their meetings with the Duke of Sussex, the sixth son of George III, who was a frequent guest. One of Coke's correspondents wrote that his visitors 'no longer felt any republican hostility to princes'. In recognition of his 'services to agriculture' and his 'steady, firm and inflexible regard for this country in its darkest hour' Coke was made an honorary member of the Massachusetts Society for Promoting Agriculture. When the Holkham tenant Mr Blomfield of Warham travelled in America, he met both George Logan and James Mease of Philadelphia, who wished he 'would settle amongst us and show us the pattern of English farming'.[19]

Through these personal contacts with the New England ruling elite, as well as through the farming literature (such as the pamphlet by Dr Edward Rigby on Holkham agriculture) that circulated in the New World and the numerous letters that crossed and re-crossed the Atlantic, Norfolk agriculture and Coke's part in publicising it was soon well known in America. The well-to-do sought introductions for themselves or their children. Mr Bainbridge recommended two young friends from Charleston to go to Holkham to see the improvement in agriculture and the valuable stock. 'They are young gentlemen who visit England for the purpose of informing themselves on the modes of agriculture pursued and to see the various breeds of cattle in the country.' Esra Weeks even sent Benjamin West, a well-known portrait painter, so that he could have a picture of Coke for his wall. 'Mr Oliver has enlarged and improved his system of farming since his return', wrote one correspondent, 'wishing no doubt to imitate your good example, so that he is now considered "Coke of Maryland".' Nothing would have given Coke more pleasure than to take his guests around his Home Farm and those of some of the nearby tenants explaining the merits of such things as Southdown sheep, rotations of crops and the various types of turnip. These North American visitors and their Whig host would have had ample opportunity to compare the ideals of the new republic, enshrined in its Constitution and operating through its Senate and Congress, with the perceived corruptness of the British court and government. Coke's hopes for parliamentary reform in Britain would surely have found a sympathetic audience.

It was not only foreign guests, disaffected royalty, the Whig aristocracy and politicians and some of the local gentry who visited for the shooting and were made welcome. A range of intellectual and cultured men also stayed for long periods, attracted both by the fine library and by Coke's outgoing and generous character. The books and manuscripts had been housed at Thomas Coke's London house until the library was completed and were moved by his widow to Holkham in the 1760s. The manuscripts were brought by land and put in a tower and linking corridor, while the printed books destined for the library came by sea to Wells. Scholars had always been welcomed in London and continued to be so at Holkham.[20]

Foremost amongst these men was the Norwich-born botanist and founder of the Linnean Society, Sir James Smith. The introduction was probably made by his maternal uncle, the Revd John Kindersley, who was a chaplain at Holkham during the dowager's day and may well have continued to hold the post under her successors. We know that Smith was at the 1788 celebrations, and by 1800 he had returned from London to live in Norwich. Here he worked on, amongst other books, his *Flora Britannica*, completed in 1804, *The Introduction to Physiological and Systematic Botany*, whose first edition was published in 1807, and his final work, *The English Flora*, which appeared in three volumes between 1824 and 1827. Unlike his Anglican clergyman uncle, he was a Unitarian and a deacon at the Norwich Octagon chapel. Both his nonconformity and his presence at the celebrations of the Glorious Revolution suggest that he was a Whig. In spite of his busy schedule, spending three months a year lecturing at the Royal Institution, he found time to visit Coke. In 1812 Smith wrote to his friend William Roscoe to persuade him to join him at Holkham. Having praised the library, where he was able to 'rummage amongst his books, drawings, manuscripts, and prints (where we every day find treasures unknown before)', Smith went on to describe his host as 'so amiable, so devoid of all selfishness with a liveliness and playfulness of manner that nobody is more entertaining'.[21]

Born in 1753, William Roscoe had spent his early adult years as a lawyer, but was always interested in the study of history and the arts, founding with friends a Society for the Encouragement of the Arts, Painting and Design in his native Liverpool in 1773. His first major work, *The Life of Lorenzo de' Medici*, was published in 1796, the year he retired as a lawyer. Three years later, he bought Allerton Hall near

Liverpool where he housed his valuable collection of books, manuscripts and art treasures. Now his interests extended beyond the literary to improving his estate. He kept all the land in hand, superintended the farming himself and began the drainage of Chat Moss. In 1812 he wrote to Coke of his efforts at marling the moss with the help of 'a moveable iron road' allowing him to travel a mile onto the moss in all seasons.[22] Roscoe's Whig background as a supporter of Fox and his interests in agriculture were such as to make him a welcome visitor to Holkham and he regularly received invitations to the sheep shearings. Like Coke, he had celebrated the centenary of the Glorious Revolution and wrote an 'Ode to the People of France' in praise of their republic. He became an MP for Liverpool in 1806, and his early speeches were in support of an honourable peace with France and called for parliamentary reform. In the small world of the eighteenth-century Westminster village Coke and Roscoe must have met. His support of anti-slavery campaigns lost him the support of the Liverpool voters and brought a swift end to a short parliamentary career.

Smith's description of the books at Holkham whetted his literary appetite. A second letter from Smith, written in the autumn of 1814, described spending 'two hours almost every day devoted to the examination of manuscripts. I am going there on Monday with our good bishop [Bathurst] … for the express purpose of looking further into these treasures', which included valuable collections of Italian documents.[23] In December 1814 Roscoe was finally persuaded to make the journey to Holkham. There he was joined by James Smith, Dr Parr and Dawson Turner. Together they were shown to the 'upper library' in one of the corner towers, where 'in consequence of their unsightly condition the collection of manuscripts and rare printed books had been deposited'.[24] While Coke's other guests were out shooting, Roscoe and his companions went through the material that may not been studied since Coke's great-uncle had brought them back from his travels. They found the experience as exhilarating as Smith had said it would be. Not only were there rare and early works from Italy, including an autographed treatise by Leonardo da Vinci and a cartoon by Raphael, but Roscoe was particularly excited to handle the book sent by Cosmo de Medici to Alphonso, King of Naples, as a peace offering and which Roscoe had mentioned in his book on Lorenzo, but had never actually seen. While these were of particular interest to Roscoe, the unpublished manuscripts by the founder of the Coke family wealth, the sixteenth-century Chief Justice Sir Edward Coke, also caused excitement. Roscoe offered to catalogue the library and supervise the binding of the loose manuscripts. Coke's genuine enthusiasm for the project shows in the letter that he wrote to Roscoe on receiving the first batch back from the binder:

> I did not delay a moment after my return in looking them over, and the
> moment I had done so I mounted to the upper library to select as many as

would fill the box, which I sent addressed to you in Liverpool, yesterday, containing thirty-two books, many of which you will find it necessary to divide into separate volumes.[25]

In reply, Roscoe sent Coke a copy of his book on Leo X, which again was received with real pleasure. In it Roscoe had written a sonnet of dedication, expressing the wish that Coke 'with benignant eye'

> *Shall o'er the leaves in pleased attention bend.*
> *Enough, if firm to truth and freedom's cause,*
> *He find thee worthy of his kind applause,*
> *And in the Author recognize the Friend.*

Coke acknowledged to Sir James Smith that it was only through him that he had met Roscoe, and in 1816 Coke went with Dr Parr and Sir James, accompanied by landowner and Merino-breeder George Tollet, to visit Roscoe in Allerton. Roscoe showed his guests around his efforts on Chat Moss and agreed to take Coke's advice on its reclamation, which involved dividing it into lots and improving one at a time rather than tackling the whole area at once.[26]

Very soon after their visit disaster struck Roscoe – the bank in which he was a partner collapsed and he was forced to sell both Allerton Hall and his own vast library and art treasures. An appeal was launched to ensure that the majority of his early Italian and north European paintings remained in Liverpool, where they now form part of the collection of the Walker Art Gallery. Coke spent £1,868 at the auction of Roscoe's collections on books and £1,105 on pictures, with a view to returning the majority of the books to their former owner. In all the auctions raised £5,150 and £5,875 respectively, so Coke's purchases made up a sizeable proportion of the total. There followed a delicate correspondence between Coke and Roscoe's Liverpool friends. Coke was anxious that Roscoe should have most of the books back, but it was thought that Roscoe would find it difficult to accept such an offer, so it was agreed that he should be 'loaned' them to help him in his studies. Meanwhile crates of books continued to go back and forth between Holkham and Liverpool. Those that needed it were rebound, but where the old bindings were salvageable, they were simply repaired. The work was carried out by the Liverpool bookbinder John Jones, who was able to employ three extra journeymen for five years to complete the work.[27] As the books were returned to him from Jones, Roscoe continued cataloguing, but the work was taking much longer and was much more difficult than predicted. At the beginning of 1820, he wrote that 'In the course of the year I hope to see the great work completed.' In 1821 Roscoe visited Holkham again to check his catalogue. It is not clear how long he was there, but on his return to his wife in Liverpool he wrote to Coke thanking him for his hospitality, 'it being the longest separation since our marriage which this day is our fortieth anniversary'.[28] 'The pleasure I have in his

[Coke's] society and that of his family', wrote Roscoe, 'has alleviated my daily labours, and has enabled me to accomplish what I should certainly not have undertaken had I been aware of the extent and difficulty.'[29] The final bills were paid for the binding in 1823 and by the autumn the catalogue was said to be ready for printing. Roscoe wanted an engraving to be made of the portrait of Coke by Lawrence as a frontispiece and the correspondence suggests that all was going well. A handsome quarto volume with illustrations prepared by Dawson Turner which Coke could give to friends was planned. A few copies would also be for sale.

At this point the project faltered. Roscoe's visit to Holkham for the final checking kept on being put off, and in October he decided that the completion of the task was beyond him, so an assistant was sought. Frederick Madden, who had been trained in the Tower record office and now worked as a cataloguer at the British Museum, was appointed at a fee of a guinea a day plus expenses. This is a clear example of the amateur, characteristic of the Georgian age and its Enlightenment traditions, being replaced by the nineteenth-century professional. Not surprisingly, Madden was very critical of Roscoe's work. 'Mr Roscoe is perfectly ignorant of the age of manuscripts and deceives himself completely in respect of Homer.'[30] Madden's first visit was in 1826, and he spent much of the first few months of 1827 copying the Holkham catalogue. Other great libraries were being catalogued at the same time, and there was no doubt some competition between their owners as to which would be published first. Madden heard in January that a catalogue of the manuscripts of the Duke of Sussex was 'on the eve of being published' and that a copy would be presented to Holkham.[31]

On 7 March 1827 Madden again made the 17-hour coach journey from London to Fakenham and on by gig to Holkham, where he found two of Coke's daughters and a granddaughter already guests. It was a beautiful day and a 'luxury to walk around the grounds and to sit in the warmth of the sun on one of the garden seats'. He was pleased to be given his old room again and he was soon back at work in the library, accompanied at intervals by Archdeacon Glover and Dawson Turner. On 16 March they finished the 'great historians'. However, a couple of days earlier Madden's concentration had begun to wander with the arrival of Coke's granddaughter Jane, now married to Coke's political ally, Lord Ellenborough. She was already notorious for her numerous affairs, including one with her cousin, George Anson. Madden was immediately smitten by her, describing her as 'not yet twenty and one of the most lovely women I ever saw. ... Lady Ellenborough is such a charmer that I find the library become a bore and am delighted to be with her and hear her play and sing to me.' He walked and rode with her and she lent him her drawing book. Finally, on 24 March he 'escorted her to her room, fool that I was. I will not add what passed!' The next day she 'pretended to be very angry with what had passed the previous night', but very soon, after a *tête-à-tête* by the lake and a sojourn in one of the hermitages there, all was made up. 'She is a most lovely and most fascinating woman.' However, the liaison was short-lived and there seems to have been no communication between them after she left a few days later. Madden returned to his work in the library, finishing the examination of all the manuscripts by 20 April.[32] The new catalogue was, however, far too large and expensive to print, and finally, in 1828, eight hand-written volumes were bound and placed in the library, but never published.

The state in which some of the manuscripts were being kept had worried Roscoe and his fellow bibliophiles. They commented on the damp and the danger that the manuscripts would deteriorate if they continued to be kept in the upper corridors. La Rochefoucauld had already described the library as 'handsome with a good collection of books' and noted that the manuscripts were 'in a sort of passage'.[33] As a result, between 1816 and 1819, new rooms were prepared. The Classical library was moved downstairs to an anteroom to the library in the family wing. Coke commissioned a portrait of Roscoe by Sir Martin Arthur Shee (now in the Walker Art Gallery, Liverpool) to hang in the room, showing him amongst his books with a bust of Fox on the table in front of him. Bookcases were built, carved and gilded for the rebound manuscripts in what had been the first earl's dressing-room.

Roscoe was not only a collector of fine art and books, a historian and critic of Classical literature, but true to the tradition of the gentleman-amateur his interests were wide and included the study of botany, which he shared with his two friends Sir James Smith and Dawson Turner, and his last book was a botanical one. He was also a poet, and his work ranged from hymns for the Unitarian church to children's verse. His works 'exemplify the muse, domestic and celebratory, of a cultured eighteenth century gentleman of a Whiggish way of thought. They help to illustrate the century's combination of classical tradition and the awakenings of interest in human rights and emancipation'[34] – no wonder he was a friend of Coke. He died in Liverpool in 1831, and perhaps his most lasting legacy to his native city was as co-founder of the Liverpool Royal Institution set up as an adult education centre in 1814. This later became part of the University College, which in its turn became the University of Liverpool.

One of his poems describes his stay at Holkham, with its 'marble halls and crested towers' and the hours spent lost in the pleasures of discovering new treasures in the library.

> *Sunk in learning's calm retreat,*
> *Midst scenes remote from vulgar eyes,*
> *I trace the weakness of the great,*
> *And mark the follies of the wise.*
>
> *But happier far the moments fly,*
> *When resting from my lengthened toil,*
> *I meet with Coke's benignant eye,*
> *And share his kind approving smile.*[35]

Here we see an entirely different side to Coke, whose cultural interests show him in a new light. He is not to be dismissed as a bucolic and self-opinionated wealthy squire with a passion for field sports. His genuine friendship and admiration for Roscoe showed not only in the regular gifts of game sent north, but also in the far more telling concern for the loss of his friend's library. Coke had a more sensitive side that understood the value of his inheritance and ensured that it could be studied by those who would appreciate it. This led to his gathering around him a group of friends who not only shared his liberal values, but were likely to prefer to spend their time at Holkham in the library rather than out with a gun. Coke's great mentor Fox had strong intellectual interests, and no doubt, particularly in the years 1794–1801 when he withdrew from parliamentary life, he too would have taken pleasure in the Holkham library and its contents, as well as in field and gun.

Such friends included the outspoken Whig, Samuel Parr. Regarded by some as 'the greatest scholar of his age' and the 'Whig Dr Johnson', he was headmaster of Norwich School from 1778 to 1785, which is when Coke probably met him. Educated at Harrow, which by the 1760s was patronised by the leading Whig families, he then went on to Cambridge before returning to his old school as a teacher. He was renowned for his Classical learning and, like Roscoe and Smith, amassed a huge library. It was while at Norwich that he wrote *A Discourse on Education*. In a letter to Parr thanking him for a copy of the book, Jane Coke wrote that she hoped that he would come to Holkham to 'judge how far I have been able to put [your ideas] into practice with respect to my children'.[36] In 1785 he left Norwich to become a parish priest at Hatton in Warwickshire, where he continued to take in pupils, including some who remained devoted to him and later became well-known scholars. The last of his pupils was Coke's nephew and heir presumptive, William. From his humble parsonage Parr would set out to be entertained as an honoured guest in the great Whig houses. He remained devoted to his little parish, where he was a model priest, but because of his Whig politics and close friendship with Unitarians such as Joseph Priestley he never rose within the Church. In 1808 Coke offered him the living of Buckingham, worth £300 per annum, of which he had the gift, but as this would require Parr to move from Hatton, he refused. However, from that date he always addressed Coke in correspondence as 'dear and honoured friend and patron'.[37] Coke and Parr shared many political beliefs, and Parr, like Coke, regarded Fox as a political hero. After Fox's death in 1806 he wrote to Coke praising Fox's achievements in glowing terms in a long letter which he later published in his two-volume work *Characters of Charles James Fox*, dedicated to Coke.

Clearly an eccentric, with a brilliant mind and renowned for his untidy dress and appearance which led to his being compared to Samuel Johnson, Parr's surviving letters at Holkham are lively and show a cheerful good humour, jumping from family matters to Classical subjects, to whether Coke has read a book about

Napoleon, to complaints about the corruption of parliament. In 1790 Coke responded to his request for £50 to support a poor scholar at Cambridge and in 1821 gave £20 to Hatton Church. In an undated letter he thanked his host for a pleasant stay at Holkham, 'Where am I more at ease – where do I meet more splendour mixed with comfort? Nowhere.' A further thank-you letter survives from 1817 recording the 'kind reception' received in 'your hospitable and noble mansion … Amidst all their diversities of situation and character, the gentlemen who visited you were, one and all, agreeable.' In 1819 he made the journey from Warwickshire, describing the road from Peterborough to Kings Lynn as 'very fine' but the country 'very dreary'. He visited again in 1821 to meet William Roscoe and the Duke of Sussex, and wrote, 'The enjoyment of Holkham (good sense, good manners, good temper, good politics) far exceed all the bustling and transient pleasure which the coronation [of George IV] will afford. I detest the tyrant and his vassals more and more.' This abhorrence of the monarchy is clear in much of his writing, as is his continued support of the French Revolution and Napoleon. In a letter to Coke written on 15 March 1816 he laid down his political creed. He supported Napoleon against 'the pilferers of his pension and the kidnappers of his person', and the army of the people of France against 'any and every foreign power which shall presume to oppose their sacred right to choose their own sovereign'. Coke would certainly have agreed with Parr's support of 'the interest of agriculture and commerce' against the 'rapacity of contractors and stock jobbers' and 'wise men' against 'blundering ministers'. In spite of his strong anti-Tory views he was able to make some light-hearted comments: 'My surgeon is a Tory', he wrote in 1820, 'but he is kept in order by my two physicians who are Whigs.'[38] Although Parr was a minister of the Church of England he supported Catholic emancipation and had a great deal of sympathy for the beliefs of Unitarians such as Smith and Roscoe. When a Birmingham mob, stirred up against the Dissenters and revolutionary societies and shouting 'Church and King' burned Joseph Priestley's house in Birmingham Parr feared he would be next. Mrs Parr and her daughters moved out of the rectory and his library was stored in a nearby barn, but Hatton remained quiet. A portrait of Parr, commissioned by Coke from John Opie in 1807, remains at Holkham.

Another member of this group of Holkham Whig intellectuals was Bishop Bathurst. Coke probably first met Henry Bathurst through his wife's family, as one of her sisters was married to a close friend of his. Although they may well have met before he became bishop of Norwich in 1805, it was not until then that he was a regular visitor. He was a good friend and strong supporter of Coke, and certainly enjoyed the company, the library, the shooting and a game of whist with Coke and his guests. He was well-read and would have been quite at home conversing with scholars as they searched through the dusty boxes in the upper corridors of the house. He was always on friendly terms with Protestant Dissenters and was well known for his support of the Catholic cause. In 1828, already an old man and in the year before Catholic emancipation was finally granted, he wrote from Bath 'the lamp is going out, but it burns clear to the last, and if, before it is quite extinguished, I should see the establishment of civil and religious liberty, I shall depart in peace. Unless desired to present a petition to the House of Lords on behalf of the much injured Catholics, I may probably never again take my seat in the House.' In fact he was there in 1832 and was the only bishop to vote for parliamentary reform. Again in 1835 he intended going to the House of Lords to take his seat in 'support of those principles to which, after your example, I have been uniformly and warmly attached through a very long life'.[39] Like many clerics of his time, he spent only a little of his time in his diocese and in the administration of its affairs. He died in 1837 at the age of 93.

It was not only the literary, but also the artistic who received a warm welcome at Holkham. Gainsborough spent time there, painting a portrait of Coke and giving his daughters lessons. A self-portrait, still at Holkham, was said to have been painted there. John Opie also worked at Holkham, and while there may have met on a visit to Norwich, the novelist and daughter of a Norwich doctor, Amelia Alderson, whom he married in 1798. Amelia was already spending time in London and was friendly with such radicals as William Godwin who were also known to John, so they may equally well have first met there. The Aldersons, like James Smith, were Unitarians who worshipped at Norwich's Octagon chapel. They were supporters of the Whigs and radical reform. Amelia's first novel, *The Dangers of Coquetry*, was published anonymously in 1790. She also published several poems in the Norwich radical periodical *The Cabinet* and was a great admirer of Fox and all that the French Revolution stood for. Her family connection with the Norwich Whigs and her personal political involvement make it likely that she would have met Coke, particularly at election time, but when, in 1800, Amelia wrote a note of sympathy to Coke on the death of his wife Jane, she admitted that her 'acquaintance with your family was a very slight one'. Coke commissioned John Opie to paint full-length portraits of Fox and Dr Parr and provided an introduction to Fox when John and Amelia went to Paris in 1802.[40] It was chiefly to see the paintings that Napoleon had collected during his conquests in Europe, which were on display in the Louvre, that John Opie wanted to visit Paris, while Amelia was more interested in watching Napoleon pass by from the window. John died in 1809, and Amelia moved back to Norwich from London, where she regularly received a gift of game from Holkham. She returned to Paris in 1829 and wrote to Coke of her meeting there with Lafayette, who again expressed his admiration for the farming at Holkham and hoped one day to visit. Amelia was one of many who wrote congratulating Coke when he was ennobled in 1837.[41]

Members of the royal family were also attracted to Holkham. Coke could never forgive George III for declaring war on the American colonies and was openly critical of the corruption of the court party and its interference in politics. This antagonism towards the king would automatically endear him to the king's eldest son, the Prince of Wales (later George IV), whose relationship with his father was always strained. In defiance of the king, the prince chose to champion the Whigs and their leader, Fox. He was a frequent visitor to Holkham and enjoyed the shooting there. However, his relations with the Whigs cooled in 1787 after his secret marriage to Mrs Fitzherbert and the embarrassment this caused Fox and his friends. Although he was invited to the 1788 celebrations and still visited Holkham occasionally, he was less warmly received than previously. In 1809 the situation changed when the Prince of Wales was declared regent in the place of his 81-year-old mentally ill father. The Prince had been the great hope of advancement for the Whig party, but his elevation to the regency heralded the completion of his political turnabout, partly because he feared the reaction from his father, should he recover his sanity. From being liberal and reform-minded he became conservative, even reactionary, and any hope Coke and his political friends had had of royal support was dashed. Contact between the regent and Holkham ceased.

This change of heart did not extend to other sons of George such as Augustus, Duke of Sussex, who became a regular visitor. His secret marriage had caused a rift with his family and he was never offered any official position, and his support of radical views further estranged him from his father and the court. He devoted his time to supporting such causes as Catholic emancipation and the reform of parliament, frequently speaking on these subjects in the House of Lords.[42] He was drawn to Holkham not only by its politics, but as the only member of the royal family with intellectual tastes he was also attracted by the library. He himself collected over 50,000 volumes as well as many ancient manuscripts. In 1822 his visit coincided with that of Dawson Turner, who wrote to William Gunn, 'His Royal Highness entirely answered your description of a most agreeable man ... he was also a very kind one.' At this time he was collecting every edition of the bible for his library (by his death, he had amassed over 1,000 editions) and was working on a new translation.[43] The Holkham chaplain commented that his conversation showed 'a vast deal of reading, and but for his appalling voice [possibly caused by his asthma] would be very pleasing'.[44] He was one of the group around Roscoe who were unpacking the Holkham books from the trunks in the upper corridors and towers.

Having no country seat of his own, the duke spent long periods staying with friends. Mrs Stirling recounts that he journeyed in a travelling coach with a large amount of luggage, and he drove at a tremendous speed with four post-horses even through the toll bars, from which, as royalty, he was exempt.[45] This description is substantiated by Thomas Creevey, who was staying at Cantley in 1822 when the duke arrived. 'This Royalty is the very devil ... Sussex arrived on Wednesday between three and four, himself in a very low barouche and pair, and a thundering coach behind with four horses.' On board were his staff, including a son of Lord Albemarle's, a black *valet de chambre* and two footmen clad *en militaire*.[46] At

Holkham two of the duke's great passions, shooting and books, could be satisfied in convivial company, and after 1818 he frequently spent up to two winter months there.[47]

Lady Jerningham, visiting in 1819, wrote of the duke, 'He is himself the most agreeable person: universal knowledge and universal taste for everything. ... He studies much and now talks of studying Arabic.'[48] He had always been subject to long bouts of illness, which meant he sometimes spent extended periods in his room 'without the slightest amusement, and probably asleep'.[49] Although this larger-than-life figure (he was six feet four inches tall and increasingly overweight) was always good company and kept up a fiction that he wanted to be treated without ceremony, in reality he expected the deference due to royalty to be shown and certain protocols to be observed. Ladies should always wear gloves in his presence. Although Thomas Coke always breakfasted at nine, the duke had to be served at ten.[50] He was one of very few men who took the liberty of smoking while at Holkham.[51] Coke, with his own views on monarchy, found this behaviour difficult to accept. When his second daughter, Lady Anson, who was acting as his hostess, pointed out on one occasion that the duke had announced that he would be arriving on a Sunday afternoon and that he would be affronted if no one was at home to receive him, Coke replied 'If the Duke chooses to time his arrival for a time when he knows I always go to church, he cannot complain of my not being at home to meet him.'[52] There were also times when his long stays were seen as irksome. Coke's youngest daughter, Elizabeth, felt the presence of royalty was a mixed blessing. She regarded the 'frequent and prolonged visits' of royal dukes as a 'considerable encumbrance' and looked forward to her marriage and new home, where 'we are not dependant on the breath of courts and princes and not bound to receive royal visits'.[53] In later years fellow guests at Holkham also began to tire of his company. 'It is not concealed that everybody thinks his Royal Highness a *bore*', wrote Napier in 1841.[54] Again this view was voiced by Creevey, who wrote, 'He has every appearance of being a good-natured man ... but there is a *nothingness* about him that is to the last degree fatiguing.'[55]

Like Coke, the duke married a second time (and again illegally, flouting the terms of the Royal Marriage Act). Society was divided as to whether to accept Lady Cecilia Underwood, later grudgingly given the title of Duchess of Inverness, as the legitimate Duchess of Sussex. Coke was one of those who knew that the marriage had indeed taken place, as it had been performed by one of the Holkham Whig literary circle, Archdeacon Glover, in 1831, and so at Holkham they were readily accepted as man and wife.[56]

In 1820 George Keppel, later Earl of Albemarle, was appointed as an equerry to the Duke of Sussex and as such frequently accompanied him to Holkham. He described him as 'among the best natured of men and best-instructed of princes' who was a consistent supporter of human rights.[57] George's brother, Henry Keppel, wrote of how he always looked forward to visits to Holkham and described it as the centre of the leading Whigs of the day. In July 1821 a party assembled at Holkham to meet the radical MP and Derbyshire landowner Sir Francis Burdett on his release from Marshalsea prison. He had been found guilty of seditious libel for an article condemning the Peterloo massacre in which he had described the king and those involved as 'bloody Neroes'. From Holkham he was escorted on a triumphal ride to the Albemarle seat at Quidenham.[58]

What all these men had in common was their 'patriotism' in the eighteenth century sense of the word. Indeed the word appeared in several of Dr Parr's obituaries. They supported what they saw as the basic British values of liberty and tolerance wherever it was to be found, whether in America or France, and abhorred corruption, particularly in the courts of kings or in government. They approved the sentiments of the Whig motto, 'Live and let live', and so supported the abolition of the slave trade and slavery, religious toleration and the ultimate goal of the reform of parliament. As these men, representing both the new and the old elite, sat together late into the dark winter evenings playing cards after a day spent out with a gun, or riding over the park and neighbouring farms admiring the fields and livestock or poring over manuscripts in the upper library, they could largely agree about politics and the changes that they would like to see. They continued to be drawn to Holkham by the spontaneous warmth and hospitality of their host and the congenial surroundings, and also by the knowledge that there would be like-minded companions there with whom to pass an evening.

Chapter Eight

SECOND MARRIAGE AND LATER YEARS, 1800–1842

The year 1800 had marked a tragic change in Coke's domestic life. One consequence of the dual loss of his wife and his son-in-law was that his daughter, Lady Andover, spent more time at Holkham as hostess. Entertaining could continue on as lavish a scale as ever. However, her remarriage in 1806 to the hero of the Battle of St Vincent, Admiral Digby, would have taken her from Holkham again. In 1818 the death of Lord Anson, the husband of his second daughter, Anne, meant that she too was frequently at Holkham, no doubt with all or some of her 11 children. Shoots continued throughout the winter until the end of Coke's life. For three successive months, Coke kept open house for his friends, with large numbers of guests entertained for weeks at a time. The future prime minister, Lord Grey, visited in 1813 and was 'delighted with the place'. The house and its treasures were 'magnificent'. He was impressed by the 'extent of the grounds and the excellent management of the farm and plantations'.[1] Battues (where the birds were driven to one place and then forced to fly high over hedges or trees towards the guns) began the first Wednesday in November, with two a week for the rest of the season. Non-battue days were spent in the turnip fields amongst the partridges or in the salt marshes going after snipe and wild fowl.[2] As Coke began to take less interest in political affairs he was able to spend more of the shooting season at Holkham.

The year 1822 saw major alterations in Coke's personal life. At the age of 68 and after 21 years as a widower he astonished his friends and London society by suddenly marrying his 18-year-old godchild, Lady Anne Keppel, daughter of his close friend, Lord Albemarle. Along with her brothers, she had been a frequent visitor to Holkham and a young companion for Elizabeth Coke. As Coke had no sons, it seemed clear that his estate would be inherited by his nephew, William Coke, and he felt that Anne would make William a suitable wife. However, neither of the young people could be persuaded to be interested in one another. William was described as a fine, generous and fearless young man with a violent temper, who, while at Eton, had swum the Thames with a hare in his mouth. His wish to join the army had been thwarted because he was the heir presumptive to Holkham.[3] Lord Albemarle then surprised his daughter by announcing his intention to marry

Charlotte Hunloke, Coke's 45-year-old niece, and Anne feared her position as her father's hostess and confidante would be lost. How long Coke had been contemplating Anne as a wife for himself rather than his nephew is not known. One story suggests that when William refused to consider the marriage, Coke replied 'By God, if you won't marry her, I will marry her herself.'[4] What is clear is that shortly after Lord Albemarle's wedding the engagement between Anne and her godfather was announced. The news was greeted with incredulity in London and Norfolk. Holland House, the home of the leading Whig, nephew of Fox and friend of Coke, Lord Holland, was not only a brilliant centre of political and literary society but also a hotbed of gossip. Here 'Mr Coke's absurd marriage to Lady Anne Keppel, fifty years younger than himself, is the general topic of conversation.'[5] Put another way 'Old Coke, in a breezy love-fit married Lady Anne Keppel, a girl of 18.'[6]Another story tells how William, realising that the marriage might well result in his uncle producing a son and thus his own loss of the inheritance, rushed to London to propose to Anne, only to find he was too late.[7] The marriage took place in London on 26 February, a fortnight after that of Lord Albemarle. The couple remained in the capital for several weeks entertaining their wide circle of friends. Coke attended parliament, where he continued to grumble about government control of the voting by packing the House with placemen, to which Lord Londonderry replied with, one imagines, something of a smile, that 'after what had recently passed immediately affecting the domestic happiness of the honourable Member, he had hoped to find him in a better temper.'[8]

The couple were at Holkham by the end of March, and were visited by friends during the summer, who reported on their happiness and the ability of Lady Anne as a hostess who could make conversation and put her guests at their ease. (This view, as we shall see, was very different from that of Alexander Napier some 20 years later.) Things were not always so enjoyable, however. One evening in October 1822 dinner was described as dreadfully glum. Coke had a cold, Lady Anne had a bad headache (she would have been heavily pregnant at the time) and Viscount Titchfield was cross with William Coke, who had shot badly that day.[9]

There were two people whose futures were directly affected by the wedding. Firstly the birth of a son at the end of the year meant Coke's nephew, William, was no longer the heir to Holkham. However, he appears to have shown no resentment at his changed circumstances. Napier described him in 1841 as taking 'his hard lot like a philosopher; an idle clever man, the life and soul of the battue work at Holkham'.[10] He never married, but moved to Norfolk, where he indulged his love of hunting until his death.[11]

Secondly there was Coke's youngest daughter, Elizabeth, who at 27 found herself in the same position in her home as Anne Keppel had feared she would be at Quidenham. Indeed, society gossip claimed that after Coke's remarriage, 'his daughter, unable to bear the changed conditions of her former home, went on a tour

Plate 10: Portrait by George Haydon of Coke's second wife, Anne Keppel (1803–1844). A daughter of the Earl of Albemarle, she was 18 when she married the 68-year old Coke, but survived him by only two years. She succeeded where Coke's first wife had failed in that she produced a male heir for the estate.

of Scotland with her sister.'[12] While staying at Dalmeny she met John Spencer Stanhope of Cannon Hall, Yorkshire, to whom she had already been introduced at Sir Thomas Beevor's Norfolk home in Hethel. They spent time together and he recounted tales of his travels to her. John was obviously viewing her as a future wife when he wrote to his mother, 'She is not so handsome as I thought and beginning to look a little older. She is very quiet, perfectly ladylike, has evidently a good deal of taste and very well inclined to laugh at the whiggery, farming and shooting of her native county.' In spite of this, she preferred the country to London and adored her father. She was bitter over his recent marriage, although she was still on friendly terms with his new wife.[13] In time their engagement was announced and the two then travelled back to Holkham, where John Stanhope first met her father. There was only a small house party when they arrived, just two of the Spencers, Sir James Smith, Lady Hervey and her sister Elizabeth Caton.[14] Stanhope's main drawback was the fact that he was a Tory and at Holkham he found himself amongst the greatest Whigs in the country. However, Coke accepted this and is said to have remarked that 'There is only one good Tory and that is Stanhope.' He was accepted as 'the son-in-law of the house, whatever his politics might be'.[15] After his initial meeting with his future father-in-law John Stanhope returned to his home at Cannon Hall, and Elizabeth wrote to him nearly every day describing life at Holkham.

Some aspects of life there were bound to be different under the new mistress, as noted by Elizabeth in her letters. The new butler had a bright chestnut wig. 'The establishment is altogether quaint and just as I had expected. I should have liked you to see it at former times ... but I am too thankful to escape all the plague I have had for many years.' Again, on relinquishing her duties as head of her father's household she wrote, 'I really feel like a captive set free; it was such a task and charge.'[16] As autumn approached the size of the shooting parties increased. On 6 November nearly 800 head were killed. 'George Anson killed 130 with his own gun.' On the 7th Lord Althorp killed 100 head. He described it as 'the best day of my life'. On 12 November the Duke of Gloucester was due with a party of 26. As Stanhope had already noted, Elizabeth had little time for farming, shooting or parliament. She advised her future husband to 'have as little to do with farming as possible ... it is a most expensive amusement', and 'What ever you do, keep out of parliament.'[17] Elizabeth and John Stanhope's wedding took place on 5 December 1822, and payments for the ringing of church bells and for horses 'to take the company from the Hall to the Church' are recorded in the household accounts.[18] The couple then moved north to Cannon Hall. Elizabeth had dug up plants from her garden to take with her, and looked forward to a quieter life there.

The new mistress of Holkham made changes to the grounds. From 1825 'florists' were employed to work in the flower garden and pleasure grounds, and from 1826 shrubs, fruit and forest trees were regularly purchased; all an indication of changing fashions in design as spearheaded by John Claudius Loudon. Gardens, as distinct from parks, grew steadily in variety, scale and sophistication. In 1831 a hot-water system was installed to heat a new greenhouse, a new flower garden laid out and a plantation created in an extension of the park on to what had been Wells Heath to the east. A new lodge had been built in 1828. It has not been possible to locate the position of the pleasure grounds, although they were probably immediately to the east of the house. In 1834 thatchers were employed to repair the roofs of the summer-houses in the pleasure grounds and a pump was installed. In 1835 the five artificial stone vases ordered by Mr Chantrey for Lady Anne's flower garden arrived.

Very little work was undertaken in the house. Small items of expenditure included the repair of furniture, particularly the rush seats on the chairs. In 1828 the large sum of £143 was spent on satin for the north state bed and for hanging the state closet. Two years later a shower-bath was installed, and in 1832 a bookcase and chair were bought for Lady Anne's room. The household accounts are sadly unforthcoming about other items – tantalisingly, this is all they record for the period.

In many ways life went on much as before, with time divided between London and Holkham, as well as visits to family and friends. In most years in the 1820s and 30s May and June were spent in London. The household accounts show board wages paid to the servants left at Holkham during this period and London expenses amounting to between £1,700 and £2,000. After groceries the major expense (about £200) was for wax candles. At a time of year with long, light evenings this suggests a lively nightlife of entertaining, and in several years the London accounts include the purchase of up to nine dozen packs of cards. Subscriptions to the Whig establishments of both Brooke's and Boodle's were kept up.[19] Accounts of the dinner-parties attended by prominent Whigs as well as members of the artistic, literary and scientific worlds abound in the journals of the period. Coke rented different houses in London over the years, and it is not clear where he was living at any one time. At the time of his marriage Coke was living in Paddington, while in 1830 he was in Kensington.

Once back at Holkham, there were agricultural matters to be dealt with. The sheep shearings ended in 1821, but there was still much to be supervised. As well as the shearing of sheep and the sale of wool, probably at Thetford wool fair, there was haymaking and harvest, with the shooting season following on from October. All had then to be ready for the major season of entertaining. The servants' liveries and the keepers' jackets might need replacing. In 1823 two silver knots for livery coats cost £4.

New books for the library would be purchased, including antiquarian editions (Roscoe was paid three guineas for a manuscript he had bought for Coke in London), new religious publications, and a three-volume history of Northampton-shire and a history of Corpus Christi College. A late addition to the library was a gift from the Norwich printer R.N. Bacon of the four volumes of the first copy of the Roman Catholic Breviary to be published in England in 1830 in appreciation of

Coke's support for Catholic emancipation.[20] In 1806 Coke bought Bloomfield's *A History of Norfolk* and in 1812 he subscribed to Richard's *A History of Lynn*, published in 20 instalments. A more surprising purchase was a copy of Finche's *Philosophy* in 1813, and in 1825 Lord Hastings wrote to Coke to let him know that the Persian dictionaries were being sent from Donnington to Holkham 'in a tin box'.[21] Lighter works to amuse family and guests included volumes in a series of Delphin classics. A copy of John Wood's plan of Lynn and of Bryant's large-scale map showing the roads and toll bars were bought for those planning longer expeditions across the county. A new mahogany stand for the library was bought in 1812. Subscriptions to the *Norfolk Mercury, Bury Post, County Times* and *Farmers' Journal*, as well as the Burnham circulating library and the County and City Reading Rooms in Norwich, were renewed. From 1801 Coke was paying a subscription to Wells Book Club for himself and his daughter.

While shooting might occupy most of the autumn day, other diversions were needed for the long winter evenings. As well as the endless games of cards for which several dozen packs were regularly purchased, other amusements were sometimes arranged. In 1823 £5 was paid to Indian jugglers and the same sum in 1827 to John Barry, a strolling player, for an evening performance. The subscription to the Kings Lynn Assembly was always renewed. There was an annual trip for the servants to the Fisher Theatre Company in Wells, with the family and guests occupying the boxes.

However, Holkham life did not suit all Coke's guests. Thomas Creevey retired from politics in 1825 and became better known as a society figure and gossip whose writings showed an acute eye for absurdity. In April 1827 he visited Holkham and was bored. There were no London papers until four in the afternoon. He was 'delighted with everything about the house, except the company who, God knows are rum enough and totally unworthy of Chief Justice Coke in creating the estate, and the Earl of Leicester in building the house. Our worthy King Tom is decidedly the best.' However, he wrote that he came to see the place and not the people. The hall, its furnishings and its pictures impressed him. Of the people, Coke was certainly the best company; Admiral Digby and Lady Andover had nothing to say, the Stanhopes were 'worthy, honest, absurd and lackadaisical'. The other guests consisted of a 'young British Museum artist [Madden] classifying manuscripts and a silent person without a name'. 'Lord John Russell and other great and public men are expected but I have known many who are expected who never come.'[22] By 1838, his opinion of the family had mellowed when he visited Holkham shortly before his death at the age of 70. This time he described Elizabeth Stanhope as 'full of good sense and a most accurate observer of persons'.[23]

Anne Keppel seems to have adjusted easily to her new role, and in spite of producing regular additions to the Coke family (five children were born in the first ten years of their marriage), the visits and social life continued as before. She attempted to bring her husband's appearance up to date by persuading him to cut his pig-tail and give up wearing a wig, but 'he powdered his head and kept his knee breeches to the last.'[24]

After 1822 there were children in the nursery again. The birth of Coke's first son, christened Thomas, at the end of 1822 (referred to by Samuel Parr in his letters to Coke as 'the young Whig' and by Creevey as 'the crown prince') was followed by that of a second, Edward, in 1824, Henry in 1827 and Wenman in 1828. A daughter, Margaret, was born in 1832, but Francis, born in 1835, lived less than a year. The Duke of Sussex was a much-loved godfather to Tom, playing with the toddler and sending excellent Christmas presents.[25] Coke's grandchildren too enjoyed the jokes and playfulness of this large royal, who was a great favourite with them. The Holkham steward, Blaikie, also took pleasure in the youngsters, describing them as 'the finest children ever'.[26]

Although Anne seems to have charmed all her visitors, her treatment of the servant class was not always as considerate. Mary Humphreys applied for the post of wet-nurse for young Henry, and believing she had been offered the post, she weaned her own child and provided herself with the necessary clothes for service in a grand house. She made the journey to Holkham several times on foot to confirm the arrangements only to be told that Lady Anne could not see her. Finally she learned that after all she was not to have the position and was offered what she saw as a very small sum in compensation. She claimed that the expense she had been put to in preparation for the post had resulted in the family getting into debt and her husband ending up in a debtors' prison. However, Lady Anne's rather high-handed behaviour was probably no different from that of many of her class, isolated as they were from the problems of ordinary people.[27]

Elizabeth Stanhope's family was also growing, with the birth of Walter in 1827, followed by a second son and four daughters, all of whom joined Coke's new family in the school-room on their frequent visits to Holkham. In 1829 William Gamble was paid for 'one year's tuition of the young gentleman' and a governess, Antoinette Baton (probably French, as was fashionable at the time), is mentioned in the household accounts in 1828. She was followed by Sophia Lulmen in 1835. The children would not have been allowed to impinge greatly on the life of the hall, being confined to the nursery. Almost inevitably there was a certain rivalry between the Coke and Stanhope nannies. Coke, however, visited the nursery at meal times, and 'would sit down and eat roast potato with butter and rock salt'. The children then went down to join their parents for dessert, after which they played games in the drawing-room with Lady Anne. Day-time activities included trips to the beach to collect shells. Elizabeth's eldest daughter, Ann,a remembered a servant's ball in the audit–room, where a band, led by the hairdresser from Wells, played.[28] Creevey too described a servants' ball at Christmas 1837. Coke, aged 85, opened the ball. 'He is a marvellous man, but I think he is *going out* tho' he burns as bright as bright until the last.' There were 25 couples, and Creevey's partner was the now-widowed Anne

Anson.[29] A few days later there was 'another great ball', and during the day Coke took Creevey for a drive 'to show me his improvements'.[30]

Elizabeth Stanhope never lost her affection for Holkham, and the family spent long periods there during the winter. Occasionally they stayed at Quarles Farm, just outside the park, which was taken in hand in 1823, but Elizabeth did not like it, writing that it 'was like being just outside the gates of paradise'.[31] While in later years she visited with her husband as the climate was said to suit his delicate health, in the 1820s she often came without him and her letters give an insight to life at Holkham at the time. She refers to her father as 'Majesty' and describes his complaints about the servants, 'who snorted at him as if they had four nostrils and blew all the powder out of his hair'. He confided to her his dislike of parliament and also of music, caused by his increasing deafness. He was one of the stewards of the three-day Norwich Festival, with concerts morning and evening. 'One day will satisfy me. I shall hope to escape for I have no taste for music.' Instead he preferred to spend a day outside. In March 1824 'he was planting all day without his great coat and reduced to brandy and water in the evening.' He was a frequent visitor to Cannon Hall, where he was unimpressed by the extensive method of keeping sheep in Yorkshire.[32]

Coke's second marriage coincided with a period of agricultural depression and fears that income from the Dungeness lighthouse would disappear as agitation against private lighthouses peaked. Unlike most other great proprietors, Coke relied almost entirely on these sources of income and the estate was encumbered with heavy debts inherited from both his father and great-uncle. During the Napoleonic Wars rents had been steadily rising in tandem with cereal prices, reaching a peak of £32,186 in 1821. Although, compared with others, the reductions on this well managed estate were small during the depression, rents never dropping below £30,000, they did not reach their 1821 level again until 1834. This prompted Coke's ever-attentive steward Francis Blaikie to take a careful stock of his employer's financial affairs. He feared that the agricultural depression, and thus the decline in rental income, might be permanent. He foresaw that there might well be other children who would have to be provided for. His frustration at not being able to persuade Coke to take the situation seriously enough was vented in a letter to the London lawyer for the estates, Mr Hanrott: 'I am quite appalled with the present and future prospects for Mr Coke's affairs. Would to God Mr Coke could see those matters in the serious light that I do.'[33] As part of this exercise, he produced in October 1822 an analysis of 'Expenses considered to be the domestic Establishment'.[34] By far the most expensive entry was £2,620 for a strange assortment of household items all lumped together but considered essential ('servants' liveries, coals, oil, wax and tallow candles, vinegar, linen, turnery, perfumery, etc.'). Next there was £2,050 for 'servants' wages, travelling expenses, carriage, postage, etc.' Housekeeping in London, plus the rent of a house there, came to just over £672.

Total expenditure came to about £16,000. It is not clear how this sum was calculated, as it does not correspond to that in the volumes of household accounts. Here, expenses of between £20,000 and £26,000 are recorded annually for the last 20 years of Coke's life. It is a figure which compares favourably with the expenditure of other great landlords. Both the Duke of Bedford, who, admittedly, had four houses to keep up, and the Duke of Buckingham and Chandos, the owner of Stowe, were spending about £35,000 a year.[35] Coke was persuaded to sell his Buckinghamshire estates, which allowed for the paying of some of the debts on the settled estates. In the event the Dungeness income was not lost for another 14 years. Blaikie's pessimism over the future profits of farming also proved to be unfounde,d and rents began a continuous and steep rise after 1834, which meant that by Coke's death, the family fortunes were sound, based on a rich and flourishing agricultural estate. However, the financial worries of the earlier years account to some extent for the lack of grand architectural schemes undertaken in the house.

The visits of the Duke of Sussex were often timed to coincide with annual dinners held in January, ostensibly to celebrate the birthday of Charles James Fox, in Norwich, but in reality to draw attention to what the Whigs saw as the un-constitutional conduct of the Tory government. The first was held in 1819, but for the 1820 event efforts were made to give the meeting a national rather than a provincial flavour and so the Duke of Sussex was invited. It took place in St Andrew's Hall. Stoves had been lit all the previous week to warm it up and lamps suspended from chains hung between the pillars to provide light. There was a circular table at the head of the room with long tables stretching down from it, while an orchestra played at the far end. The names of Fox, the Duke of Sussex and Lord Albemarle (the chairman in 1820) were illuminated above. A bust of Fox was placed in 'a sort of temple' in the centre of the round table. It was surmounted by a 'figure of Fame' holding a flag of light blue (a Whig colour) on which was inscribed 'Magna Carta', 'The Bill of Rights' and 'Trial by Jury'. Dinner for about 460 guests began at 5.30 p.m., and when the feast was over and the cloth removed it was time for the speeches. In 1820, it was the Earl of Albemarle who presided and toasts were drunk to 'The King', 'The Prince Regent, 'The cause of civil and religious liberty', 'May the example of one revolution prevent the necessity of another', a 'Full, fair and free representation of the people in parliament', and 'The constitution according to the principles of the Revolution', all of which were greeted by loud applause. In contrast, the toast to the memory of Chares James Fox was drunk 'in respectful silence'. In his speech Albemarle spoke of the present situation as 'something short of a revolution in the Constitution' and the discontent 'now sullenly and silently indeed prevailing'. All this could be blamed upon the government and the distress caused by over-taxation. The increase in the size of the Army in peace-time and the provisions of the Six Acts were 'as congenial to the Congress of Aix le Chapelle and the members of the Holy Alliance as they were uncongenial to free-born

Englishmen'. If Mr Fox's principles had been adhered to, this would have been unnecessary. His word were greeted with loud applause. There followed a toast by Bishop Bathurst to 'Civil and religious liberty' and then Albemarle introduced Coke – 'who needs no introduction – you all know and love him'. Amidst cheers, Coke rose to speak. Harking back to the days of Fox, he listed Pitt's broken promises over parliamentary reform, the repeal of the Test and Corporation Acts and Catholic emancipation. Although Pitt had spoken against the slave trade, it was Fox who had finally carried through the bill. Speaking of the current government under Lord Liverpool, he said that 'more mischief had been done in a short space of time than he had ever known'. Further speeches were made and toasts drunk and the celebrations finally ended at 1.30 in the morning.[36]

The dinner of 1822 was held in the Assembly Rooms and was again attended by the Duke of Sussex, whom Albemarle described as 'ever mindful of the principles which had seated his family on the throne' and a 'numerous company of Noblemen and Gentlemen of this city and county'. His invitation to toast Coke was greeted by 'thunders of applause.' Albemarle described him as 'an individual with whose long approved merits and distinguished worth they were all well acquainted ... who held to the principles of that Great Man whose birth we are celebrating'. Economic distress was still being felt both in Norwich and the county as a whole and was attributed to government mismanagement and the consequences of an unnecessary war. At Fox's birthday dinner the previous year Coke had been cheered as he bemoaned the present state of affairs and the government's attempt to 'fetter the people and render this, once free, a despotic government'.[37]

Although Coke was described inside the Assembly Rooms and among friends as someone who could not be surpassed in 'simplicity of manner, singleness of heart and the doing of duty',[38] his attitude towards agricultural protection was, as we have seen, making him unpopular with the agricultural interest at large. By the mid 1820s the Fox dinners had either come to an end or were not regarded as important enough to be reported on by the local press.

In later years Coke developed a commitment to the Masonic fraternity. Although the origins of the Freemasons can no doubt be traced back to the medieval guilds, their existence as a society separated from the craft of masons, but with its own morality enshrined in the words 'benevolence, fraternity and utility', as well as its ritual and symbolism, can be traced to early eighteenth-century London, where the Grand Lodge of England was set up in 1717. By 1730 there were a few Lodges outside London, with that at the Maids Head, Norwich, being founded in 1725 and one at Kings Lynn 1729. Thirteen more had been founded in Norfolk by 1759, when it was agreed to appoint a Grand Provincial Master. Edward Bacon, the son of a Norwich MP and who inherited Earlham Hall in 1734, was appointed. He went on to become MP for Lynn and retained a seat for most of the following 42 years. The following Grand Provincial masters are familiar names: Sir Edward Astley, Henry

Hobart, William Bulwer of Heydon, William Palgrave, Mayor of Yarmouth, and Sir Jacob Astley, who died before he could be installed in 1816. The Masonic movement soon attracted the support of royalty, and in 1790 the Prince of Wales was elected Grand Master of one of the two rival branches of the movement. In 1812 he resigned as his position as Regent meant he would be too busy to take on the responsibilities of the office, and in 1814 his brother, the Duke of Sussex, became Grand Master. He succeeded in uniting the two divergent branches and held the position for 30 years. It may have been the influence of the duke which persuaded Coke to become involved, although he was following in the footsteps of his great-uncle, Thomas Coke, who was possibly introduced to the Masons by Robert Walpole and was made Grand Master in 1731. Some 88 years later, his great-nephew agreed to the provincial appointment, so long as he had an efficient deputy to do the work: 'Mr Coke is so much engaged in Parliament, Agriculture and extensive domestic circles.'[39]

On 23 August 1819 he was installed as Provincial Grand Master in Norwich with enormous pomp. The ceremony took place behind closed doors and shuttered windows in the Assembly House before 350 brethren invited from the various Norfolk Lodges. This was followed by a procession to the cathedral and a service, after which there was banquet at St Andrew's Hall and another procession through Norwich by torchlight, all providing a splendid spectacle for the citizens. Two days later a Grand Chapter was held at Holkham.[40] Coke does not appear to have taken any further active part in the movement, but in both 1823 and 1824 he conferred with the Duke of Sussex on his visits to Holkham over the granting of warrants for the establishment of new Lodges in Lynn and both men signed the warrants for their foundation. Able deputies carried out the necessary administration, but from the 1820s to the 1860s, there was little growth in membership.[41]

Coke's retirement from politics in 1832 meant that he was free from commitments to London life and could enjoy the shooting season without calls to the House of Commons. In 1834, he wrote to Fox's nephew Lord Holland, who was a life-long friend, 'Sporting, farming, planting and building a wall around my park fills up my time ... and with a mind at ease, satisfied that every possible good will be effected and rendered to the community at large by an honest, liberal and en-lightened government, such as I never expected to see.'[42]

In June 1837 the last of the sons of George III to inherit the throne died, and with him the much discredited Georgian dynasty. The extravagance and dissolute lifestyle associated with royalty and the court and reflected in the corruption and attempted control of parliament came to an end and was replaced by a serious, more socially conscious age. The monarchy under George IV and William IV had been at its lowest ebb since Charles I but the accession of the young Queen Victoria, who was immediately popular, marked a change for the better in public life. The last five years of Coke's life were a time seen as the dawning of a new age, politically, economically and culturally. Princess Victoria had visited Holkham with her mother, the Duchess of Kent, in 1835 and had been greeted with due ceremony. Many of the complaints that Coke and his fellow Whigs had made in the past against royal interference and the support of placemen in parliament were now things of the past. The boisterous excesses of the later Hanoverians were replaced by altogether more decorous standards, and Victoria's court became a model of sober family life. Under the guidance of Lord Melbourne, the young queen came to accept the limited role of a constitutional monarch.

The most obvious change brought about by the new reign at Holkham was the elevation of Coke to the peerage. In July, 1837 Lord Melbourne, on behalf of the queen, wrote to offer him an earldom and Coke chose the title Earl of Leicester, adding to it 'of Holkham' to distinguish it from another earldom of Leicester which had been granted to the Townshends of Raynham in 1784. Thus after resisting elevation to the House of Lords throughout his parliamentary career, Coke accepted the honour. Now that he had retired from politics and had a son to inherit the estate, he was ready to give up the title of the 'greatest commoner in England' for a more enduring one. However, he took no pleasure in attending the House of Lords, which he described as 'the hospital for incurables'.[43]

In the same year Coke was ennobled, his brother, Edward, died. He had lived at Longford all his life and now it returned to Thomas, who spent increasing amounts of time there in the spring. Most of the rest of the year was spent at Holkham, and although by now an old man Coke entertained the usual crowd of guests for the shooting in the winter.[44]

In August 1841 a new household chaplain arrived at Holkham. Alexander Napier was the son of Professor Alexander Napier of Edinburgh, editor of the *Encyclopaedia Britannica* and the Whiggish *Edinburgh Review*. Alexander was 26 when he took up his position, and wrote regularly to his father and sister describing life at Holkham as it was in the last year of Coke's life. The journey from Edinburgh took him two days and he was met at Holkham by the retiring chaplain and tutor to the boys, Mr Collyer, who was moving to the living at Warham. He found the earl and Lady Leicester 'pleasant and obliging', and before dinner he had time to wander through the house and grounds. He wrote that the park was 'unrivalled' and that it would take 10 days to examine the pictures alone. Dinner was at 4.15 p.m. He described the earl as 'very infirm', with failing sight and hearing. 'He talks a great deal of the old days very whiggishly. In the evening Lady Leicester and Mr Collyer alternately read the papers to him.' The next day family prayers were at 9.15 a.m. Coke always attended, 'though he cannot hear a word – he goes for example's sake'. Lady Leicester was 'pleasant looking, amiable, dressed with simplicity in a blue silk gown with white flowers and a lace collar. She limps from her late accident.' Later in the day he went round the house with Mr Collyer, 'every room crowded with fine pictures'. At 4 p.m. they met in the statue gallery to await the announcement of dinner and found Lord Leicester resting on a sofa. The meal was 'plain and substantial', consisting of watery vegetables, soup served from the top of the table and duck from the bottom, followed by roast pig and stewed venison with 'two side dishes of cutlets' and then pudding and pie. The port and sherry were 'good' and the beer 'glorious of which the ladies liberally partook. The talk was all about eating. ... The old gentleman seems to think it very doubtful whether good can exist without Whig feelings.' After dinner there was a drive in the park and at 8 p.m. they returned for tea and the papers were read to Coke. Sometimes there was a game of whist, of which Coke was very fond. Napier would help him arrange his cards as he could not see, but 'he never made a mistake'. At 10 p.m. 'the household retires ... Punctuality is the very life and soul of the establishment.'[45] Napier was immediately popular in the household and Coke's son Henry, who was 14 when Napier arrived, later described him as 'my closest and most cherished friend'.[46]

On Napier's first evening the only guests were Lady Leicester's sister, Lady Georgina Hill, and Mr Ellice. Napier found that Lady Leicester had very little to say, while her sister was very different. 'I have heard from her some rather startling language too ugly to write.'[47] Edward Ellice was by this date an almost permanent guest at Holkham. A Whig MP for Coventry, his politics were very much in tune with those of Coke. He had been married to Hannah, the sister of Whig prime minister Lord Grey, and had been largely responsible for negotiating the Reform Bill through parliament. For a short time he was a member of the cabinet in Earl Grey's government, firstly as joint secretary to the treasury and a government whip and latterly secretary for war with a seat in the cabinet. He resigned in 1834, but continued to be influential behind the scenes, and would have kept Coke in touch with the political gossip from London and Paris, which he often visited. Napier described him as a 'most pleasing and kind man'. He ran the household for Lord Leicester, ordering the wine and dinners. Although, at 57, he was not young, Ellice probably provided more lively company for 36-year-old Lady Leicester than her 86-year-old husband, especially as he was known to be 'urbane and sociable, displaying a marked taste for female company'.[48] He was looked on as a sort of father-figure by the household, and young Henry Coke described him as the 'chief custodian of my youthful destiny ... his kindness of heart, his powers of conversation, along with his striking personality and ample wealth combined to make him popular'.[49] He arranged for Henry to spend some time with a family in France as part of his education. Almost exactly a year after Coke's death, Lady Anne married Edward Ellice, only to die in childbirth at Longford the following July at the age of 41.

Napier spent a lot of time with Coke and wrote 'the more I see of Lord Leicester, the more I admire him'.[50] His descriptions give a picture of Coke as an old man. He was increasingly deaf and losing his sight (he bought his first pair of spectacles in 1828)[51] and his handwriting became difficult to read, until, in his last years it was his wife who kept up his correspondence.[52] His main topics of conversation were farming and politics and he showed very little interest in literature or other intellectual pursuits. He had inherited his great-aunt's prejudice against higher education and none of his sons attended university. He continued to take an active interest in his estates, and together Napier and Coke went to Kipton Ash sheep fair and visited some of the tenant farms on the way.

> This was a delight. Old as he is, he became young again at mention of anything of the harvest. He is full of his Southdown sheep and Long-horn cattle. His farmhouses are mansions, and his object he says is to make his tenants happy, interested in their farms and independent of him.[53]

On another occasion Napier went for a drive with Coke and the Duke of Bedford's agent and 'it was a pleasure to see the old Earl with his tenants'.

Napier's duties as household chaplain included setting a little Latin homework for young Wenman, who was 12 and home from Rugby as the school had been closed after an outbreak of typhus. In the shooting season, the exercise was frequently left undone. Napier was allowed free access to the library and in later years became the librarian – a post inherited by his son.

In November the shooting season began and this impressed Alexander Napier no more than it did Elizabeth Stanhope. There were three gamekeepers in Holkham park in 1839, plus one in each of the neighbouring parishes of Quarles and Warham. The battues which could result in the death of 400 birds in a day were 'positively dangerous ... everyone gets hit occasionally'.[54] Nor did he think much of the company. The visitors included royalty, dukes and duchesses, members of the cabinet and MPs, many of whom had very little to say for themselves and Napier felt he had little in common with them. Their conversation was 'cliquish'. They 'eat and drink a good dinner, and fine wines and then fall asleep after dinner for want of anything better to do'.[55] Sometimes there were friends of the young Cokes and when this was the case there was dancing in the evening. The shooting-party could consist of 20 to 30 guests plus their servants, which could double the number eating in the servants' hall. He found Lord Spencer 'a most silent man, except about bulls and cows. Only touch upon that and off he goes.'[56] On Sundays he went to see the cattle rather than attend church. Coke seems to have enjoyed the shooting and the company into the last years of his life, although his failing eyesight and increasing deafness must have been something of an impediment. In 1839 Ellice wrote to Lord Holland, 'You have no conception of how the old gentleman enjoys his society. I have not seen him better for the last five years, both in mind and body – and out with the shooters from ten to sunset every day. He is gone today to the Union workhouse which he attends regularly every Wednesday, and takes a great interest in their proceedings.'[57] In 1839 Lord Anglesey wrote to Lord Holland after a fortnight's visit to Holkham that he was 'delighted to find Coke so well and so perfectly happy. Lady Anne plays her part so admirably that it is not possible to remark the disparity of years. She neither nurses him nor courts him nor neglects him.'[58]

Elizabeth Stanhope's daughter Anna remembered her last visit to Holkham during her grandfather's lifetime in 1841, when she was 17. 'Had I not spent that winter at Holkham, I never should have realised the enthusiastic way my grandfather was worshipped by those around him.' Elizabeth, meanwhile, spent time in her room drinking tea with the Duchess of Bedford and talking of old times.[59]

In January 1841 Coke attended his last audit dinner. The event in the audit room had been overseen by Mr Baker, the agent, and Coke's two younger sons, but as the proceedings were drawing to a close the old earl joined them. The local press reported the enthusiasm with which his unexpected arrival was greeted and with which toasts to him were drunk. He then embarked upon a speech in which he made it clear that he did not expect to live to see another such occasion and reiterated his belief in the importance of trust between landlord and tenant, acknowledging the tenant's importance to the realisation of his agricultural dreams. All those present who had farmed at Holkham for all their adult lives would have known no other landlord and many were said to be visibly moved by the earl's words.[60] Six months later they would be forming part of his funeral procession to the family vault at Tittleshall.

THE GENTLEMAN PATRIOT

The world in which Thomas William Coke's son, another Thomas, and now heir to the title Earl of Leicester, would operate was radically different from that of his father. The passing of the Reform Bill and the accession of the young Victoria marked the end of an era. Politically, the old Whig oligarchy based on landed wealth was broken, and new names, often supported by industrial money, began to enter politics and landed society. The young Thomas himself married Juliana Whitbread of the Bedfordshire brewing family. He took little interest in politics and rarely attended the House of Lords. However, his younger brother, Edward, kept up the family tradition and was a Whig MP for west Norfolk from 1847. A painting depicting the election shows him climbing the greasy pole a close second to the leader in the poll, the Tory William Bagge (Colour Plate 15). The implication is that he is being pushed up by the wealth and influence he was able to exert. The Tory *Norfolk Chronicle* was not slow to point this out, nor to make the most of the fact that Edward seemed to take so little pride in his achievement that he did not turn up for the chairing in the market-place, much to the annoyance of his supporters.[1] He seems to have played only a very minor role in politics at Westminster.

Agriculture too was changing. The day of the amateur with purely 'practical knowledge' so much admired by Blaikie was passing. Agricultural experiment was now the province of the professional scientist and technologist – a change which, as we have seen, was bemoaned by Coke in his letter refusing membership of the new Royal Agricultural Society with its motto 'Practice with science'.

When, very soon after his father's death, the new earl began to stamp his own mark on the buildings at Holkham it would be in a new style, influenced by the Victorian interpretation of the Gothic, rather than the Classical styles of previous generations. Formal gardens, with terraces connected by gravel paths to an oval basin, were created to the south of the house by W. A. Nesfield and William Burn, while the new stables, carriage-house and laundry reflected the architecture of the house, albeit in a far heavier way. The popular Victorian architect S. S. Teulon was commissioned to design a porch on the north front, thus protecting the great entrance hall from the northerly winds. Although intrusive, it is remarkably similar in form and material to the original house. The Gothic screen linking the almshouses at the north entrance to the park is also by Teulon, and with the arrival of that very Victorian invention, the railway, the north entrance would replace the Georgian-style approach up the south avenue as the main introduction to the hall. The old Georgian windows with their small panes and glazing bars were replaced by technologically advanced single sheets of plate glass, only to be returned to their original state a century later.

The hall itself, however, with its paintings, furnishings and treasures remained largely unchanged and would survive as the family home of the Leicesters into the twenty-first century.

Within a few days of Coke's death there were already those who were thinking in terms of erecting a permanent memorial to the county's longest-serving Member of Parliament, whose publicising of Norfolk farming during the prosperous years of the Napoleonic Wars had ensured the county's reputation as the home of 'improved agriculture'.

Not surprisingly it was one of Coke's most successful tenants, John Hudson of Castle Acre, described by a local clergyman as one of the 'princely yeomen of Norfolk', who organised a meeting at the Swan Inn in Norwich to discuss the most appropriate way to commemorate his late landlord. John Hudson had become a tenant in the depression year of 1822 and rose to become one of the most wealthy farmers in the county, renting over 1,000 acres and said to be worth £100,000. He was a founding member of the Royal Agricultural Society and later the Farmers' Club.[2] Other leading figures on the committee were James Neave of Wymondham and Richard Noverre Bacon, the editor of the Whig *Norwich Mercury*. The initial meeting was attended by 'Noblemen and Gentlemen, the largest Landowners and the most influential and independent Yeoman of the County'.[3] Robert Leaman of Whitwell Hall was elected chairman and Bacon secretary. Over a thousand subscribers contributed a total of about £5,000.

Progress was slow. The first problem was deciding what form the monument should take. The most favoured options were a statue or a column, although the Bishop of Norwich suggested a 'model farm' and some favoured an agricultural institute. It was the column that finally found favour. There was also the question of where it should stand. The committee thought it should be in a public place, such as the cattle market in Norwich or in the cathedral. After a stormy public meeting open to all subscribers and attended by a large contingent from Wells, it was agreed that the site should be decided by a postal vote, and the suggestion of a site within the park won most support.

The design of the Swaffham architect William Donthorn was selected. Reminiscent of Nelson's column in Trafalgar Square, it consisted of a fluted column 120 feet high surmounted by a wheatsheaf, which Elizabeth Stanhope thought looked like a 'vulgar evergreen flower'.[4] This was supported on a capital decorated not with acanthus leaves, as might be expected of a true Classical design, but those of the turnip, the plant which made possible the growing of more cereals on the light lands of Norfolk. The pedestal was to be adorned with bas-reliefs by the sculptor son of the better-known John Henning. They would depict Coke's achievements. These included scenes showing the sheep shearings, the granting of a lease and the digging of water meadows. Representations of agricultural implements and statues of pedigree livestock were placed at the four corners. There were constant disagreements between Donthorn and the builder. Bacon claimed to have had a significant hand in the final design, reporting that the architect had little idea of what an ox or a sheep looked like and it was he who provided the sketches for them. On 16 August 1846 the foundation stone was finally laid amidst great festivities. The monument was actually completed five years later, but did not meet with Elizabeth Stanhope's approval. Remembering fondly the old view, she wrote that the monument was 'too near and too frightful' (Colour Plate 13).[5]

As well as the statues and bas-reliefs, there were also inscriptions drawing attention to Coke's main achievements. Under the Devon ox are the words, 'Breeding in all its Branches' and under the Southdown sheep, 'Small in size, but great in value', reflecting Coke's interest in animal breeding and appreciation of the importance of sheep in the Norfolk crop rotations. The Whig motto 'Live and let live' and 'The Improvement of Agriculture', which are on the two other corners under the plough and the seed drill, reflect his political and agricultural thinking. The long inscription cast in bronze on the north side lists Coke's achievements as the promoter and benefactor of agriculture and the arts, and salute a political career marked by integrity and independence. The newspaper report describing the laying of the foundation stone describes his personal encouragement of the 'scientific, the practical and the eminent in every degree and every art ... which led this Patriot to make agriculture his chiefest study'.[6]

So how should we assess the life of Thomas William Coke of Holkham? Should he be seen as an exceptional landlord, uniquely responsible for improvements on his estates which set an example for others to follow – a man whose towering personality dominated county life and politics for a generation and whose local standing meant that his achievements on his estates would be noticed worldwide and would resonate with Norfolk farmers and in histories for generations to come? Or was he a simply good self-publicist with the largest estate in the county who was a typical, rather than outstanding member of his class?

In reality, he was a member, albeit a prominent one, of a small group of Whig grandees with a common set of goals and aspirations placing them firmly in the Hanoverian period. In politics their brand of Whiggism had been personified in Charles James Fox, whose memory they continued to revere throughout the years of Tory ascendancy. Their creed was articulated in the Bill of Rights, which limited the role of the monarch and proscribed arbitrary government. Although the returning prosperity of the later years of Lord Liverpool's Tory government meant that after the mid-1820s the Fox anniversary dinners faded from the Whig social calendar, it was not until 1828 and 1829 that two of Fox's flagship causes, Catholic emancipation and the repeal of the Test and Corporation Acts, allowing Protestant Dissenters to hold public office, finally became law. A third great Whig cause, that of limited parliamentary reform, was achieved in 1832. After that date, politics took a different direction, leaving the old Whigs behind. Coke himself joined the ranks of the aristocrats, and the old-style Whig patriotism of which he was so proud and which could encompass support of the American colonies against their despotic British master, and was frequently referred to in his obituaries, was to give way to the nationalism of the Victorian empire.

Amongst the group too were many others with an interest in agricultural improvement. Their close social circle meant that men such as Coke, the Duke of Bedford and Earl Spencer regularly visited each other's estates, and the smaller players too, such as John Curwen and George Tollet, were also accepted members. Coke's achievements at Holkham were in no way unique, but can be replicated on estates up and down the country. Nor indeed could the great landlords orchestrate a 'revolution' in agriculture on their own. It was the numerous clergymen and minor gentry, rather than the large landowners, who provided articles for Arthur Young to publish in his *Annals*. The great capitalist farmers of estates such as Holkham were a class on their own, as is evidenced in the fine Georgian farmhouses, grand enough for the lesser gentry, which were built for them on the great estates across the country. It was on their adoption of new ideas and injections of working capital that the new farming depended. What set Coke apart was his love of publicity and his ability to do everything on a grand scale. It was up to men such as Coke to provide the infrastructure and support which allowed their tenants to flourish. It was they, in their turn, who formed the backbone of the committee that set about erecting a suitable memorial to their landlord.

NOTES

PROLOGUE

[1] Holkham MS F/TWC 10.

[2] Schmidt *et al.* 2005, 46.

[3] Stirling 1908, I 445.

[4] Holkham MS A47 261, 263, 264.

[5] Thompson 1963, 79.

[6] Robinson 1988, 155

[7] Gillen 1976, 234–5.

[8] *Derby and Chesterfield Reporter*, 14 July 1842.

[9] Young 1771, I 175.

[10] Stirling 1908, II 481.

[11] *Norwich Mercury*, 14 July 1842.

[12] It may seem surprising that the coffin was laid in inns rather than the local church. Later in the century, this would certainly have seemed very inappropriate.

[13] *Derbyshire and Chesterfield Reporter*, 14 July 1842.

[14] *Norwich Mercury*, 14 July 1842.

[15] Ibid.

[16] *Derby and Chesterfield Reporter*, 14 July1842.

[17] Ibid.

[18] *Norfolk Chronicle*, 9 July1842.

[19] *Derby and Chesterfield Reporter*, 7 July1842.

[20] *Norwich Mercury*, 6 May 1838.

[21] Maxwell (ed.) 1903, II 76.

[22] *Derby and Chesterfield Reporter*, 7 July 1842.

[23] *Norfolk Chronicle*, 9 July 1842.

[24] Griffiths 2002, 1.

[25] Ibid., 21; Cunningham 1989, 57. (Samuel Johnson was a Tory).

[26] Griffiths 2002, 32.

[27] Colley 2003, 145.

[28] Griffiths 2002, 31.

CHAPTER 1

[1] *Derby and Chesterfield Reporter*, 7 July 1842.

[2] Stirling 1908, II 328.

[3] Holkham MS F/TWC 10, 36.

[4] *Norwich Mercury*, 27 April 1776.

[5] Young 1771, I 177.

[6] James 1929, 222–3.

[7] Walpole 1937, 43, footnote.

[8] Stirling 1908, I 62–6.

[9] Rubenstein 2004, 468–9.

[10] Quoted in James 1929, 292.

[11] Holkham MS F/TWC 10, 40

[12] Holkham MS Family Deeds 58.

[13] Holkham MS F/TWC 11, 2.

[14] Cust 1899, 115–16; Card 2001, 86–7.

[15] Card 2001, 96–7.

[16] Ketton Cremer 1982, 185; Rosebery 1913, I 9.

[17] Holkham MS F/TWC 10, 41–2.

[18] Ibid., 43.

[19] Stirling 1960, 19.

[20] For a detailed description of Thomas Coke's Grand Tour, see Moore 1985, 33–40.

[21] Mortlock 2007, 138.

[22] Black 1999, 103.

[23] Young 1771, **II** 2–6.

[24] Ibid., 6–9.

[25] Holkham MS F/TWC 11, 3.

[26] Holkham MS F/TWC 10, 45.

[27] La Rochefoucauld 1988, 195.

[28] Schmidt *et al.* 2005, 142.

[29] Holkham MS F/TWC 10, 46.

[30] Ibid., 47–8.

[31] Quoted in James 1929, 294.

[32] *The Norfolk Tour, or a Traveler's Pocket Companion* (1773), 14–27.

[33] Schmidt *et al.* 2005, 138. For a more detailed consideration of the building and interiors of Holkham see ibid. 29–52, 81–175.

[34] Black 1999, 18.

[35] NRO WGN 2/1.

[36] NRO WGN 3/1.

[37] Ibid.

[38] Ibid.

[39] T. Martyn, 1787 *The Gentleman's Guide in His Tour through Italy with a Correct Map and Directions for Travelling in that Country* 29.

[40] Stirling 1908, **I** 105.

[41] Timothy Mowl has argued that much if the symbolism within the architecture and decoration of the great Whig houses, including Holkham, such as the use of garlanded oak leaves, indicates that their owners had Jacobite sympathies (lecture, Saltram House, 2007). See also Clark, J. (1992) 'Palladianism and the Divine Right of Kings; Jacobite iconography' *Apollo* 4 224–249.

[42] Holkham MSS F/TWC 10, 49–53 and F/TWC 11, 4–6.

[43] Moore 1985, 67.

[44] NRO WGN 5/ 3/23.

[45] Riviere 1965, 358.

[46] Stirling, 1908, **I** 118.

[47] Coke 1896, **IV** 268–9.

[48] Stirling 1908, **I** 122.

[49] The picture is illustrated in Moore 1985, 64.

[50] Stirling 1908, **I** 122.

[51] Ibid., 125.

[52] Moore 1985, 69.

[53] Alexander Pope *The Dunciad*, 1742.

[54] Black 1999, 291, quoting Thomas Pelham's letter to his father, 1776.

[55] Ibid., 294, quoting John Villiers, 1788.

CHAPTER 2

[1] Kingley 1992, 173–5.

[2] GRO D678/1/ E1/3.

[3] Stirling 1908, **I** 144.

[4] GRO D678/1F7/1/ 9–12.

[5] Stirling 1908, **I** 143.

[6] Holkham MS F/TWC 10, 58–60.

[7] Ibid., 41

[8] Baring 1866, 189.

[9] Winkley 1986, 104.

[10] *Norfolk Chronicle*, 15 June 1790.

[11] Thorne 1986, **II** 287–96.

[12] *Gentleman's Magazine*, 1843, new series 19 316–17.

[13] Holkham MS F/ TWC 10, 61.

[14] Ibid.

[15] Jewson 1975, 3.

[16] Joshua Larwood, *Erratics*, London, 1800, 112.

[17] Holkham MS F/TWC 23, 1796 election loose papers.

[18] Holkham MS LB 1817, F/TWC/20, 1789 election loose papers.

[19] MS Diary of Thomas Moore of Warham, 1799–1811. Family History Centre, Norwich.

[20] Holkham MS F/TWC 14–18.

[21] *Gentleman's Magazine* 1840, new series 15 317.

[22] Thorne 1986, **II** 478.

[23] Wade Martins 2005, 126–7.

[24] Holkham MS F/TWC 18, 1789 election loose papers.

[25] Thorne 1986, **I** 344.

[26] Brooke 1968, **I** 142.

[27] Ibid., 162.

[28] Thorne 1986, **I** 299.

[29] Holkham MS F/TWC 10, 76.

[30] Mitchell 1992, 105.

[31] British Library Add. MSS 51468, fos 26–8.

[32] Mitchell 2005, 622.

[33] Wright 1815, **IV** 147.

[34] Russell 1854, **III** 86–93.

[35] *Norwich Mercury*, 7 August 1830.

[36] Johnson 1843, 3.

[37] Stirling 1908, **I** 170.

[38] O'Gorman 1997, 195.

[39] *Norwich Mercury*, 31 January 1778.

[40] Holkham MS F/TWC 15, loose pages.

[41] Ibid., loose pages.

[42] Holkham MS F/TWC 11, 67–8.

[43] O'Gorman 1997, 227–8.

[44] *Parliamentary Register* 6 1782, 262.

[45] Ibid., 282–3.

[46] Ibid., 343.

[47] Stirling 1908, **I** 209.

[48] Bacon 1821, 73–4.

[49] *Parliamentary Register* 6 1782, 343.

[50] *Norwich Mercury*, 9 July 1842.

[51] Quoted in Stirling 1908, 1 205 but cannot be located at Holkham.

[52] Holkham MS F/TWC 10, 88.

[53] Stirling 1908, 1 216.

[54] Ibid., 217.

[55] Holkham MS F/TWC 1, 13.

[56] Ibid., 4.

[57] *Parliamentary Register* 10 1783, 489.

[58] *Norwich Mercury*, 5 February 1780.

[59] *Parliamentary Register* 3 1781, 138.

[60] *Parliamentary Register* 13 1784, 49.

[61] *Bury Post*, 26 February 1784.

[62] *Norwich Mercury*, 27 March 1784.

[63] *Norwich Mercury*, 3 April 1784.

[64] *Norwich Mercury*, 10 April 1784.

[66] *Norwich Mercury*, 17 April 1784.

CHAPTER 3

[1] Quoted in Stirling 1908, **I** 239.

[2] Holkham MS A49 29 and 143.

[3] Maxwell 1903 **II** 111.

[4] BL Add. MSS 48 218, quoted in Lummis and Marsh 1990, 71.

[5] Mowl 2006, 64.

[6] Williamson 2005, 58–63.

[7] Baring 1866, 72.

[8] Holkham MS A48 1802.

[9] Holkham MS A46 1782 and 1783.

[10] Holkham MS E/W, 1.

[11] Farington 1978, **II** 605.

[12] Wade Martins 2002, 62.

[13] J. C. Loudon, *The Landscape Gardening and Landscape Architecture of the Late Humphrey Repton Esq.*, 1840, quoted by Gore and Carter 2005, 17–18.

[14] Holkham MS Repton's 'Red Book'.

[15] Gore and Carter 2005, 26.

[16] Farington 1979, **IV** 1348.

[17] Holkham MS F/TWC 10, 2.

[18] Curwen 1809, 238.

[19] Farington 1982, **VIII** 3130.

[20] Brander 1964, 87–101.

[21] P. Delabere Blaine, *An Encyclopedia of Rural Sports*, 1838, 856, quoted by Williamson 1995, 138.

[22] Williamson 1995, 138–9.

[23] W. O. Hassall MS notes, Holkham Archive.

[24] Farington 1978, **II** 384.

[25] MS Diary of Thomas Moore of Warham. Norfolk Family History Centre, Norwich.

[26] Stirling 1913, **II** 48.

[27] Farington 1978, **III** 929.

[28] Russell 1853–7, **III** 89.

[29] MS Diary of Thomas Moore.

[30] Ibid.

[31] Stirling 1908, **I** 441–2.

[32] Stirling 1913, **II** 48 and 75.

[33] Holkham MS Game Books.

[34] Ilchester 1937, 176

[35] Baring 1866, 190.

[36] W. O. Hassall MS notes.

[37] Farington 1982, 8 3049–51.

[38] Parker 1975, 135.

[39] Stirling 1908, **I** 325.

[40] Young 1768, 8.

[41] Hiskey 2005, 178.

[42] Gore 1963, 436

[43] Baring 1866, 258.

[44] Holkham MS Typescript inserted in F/TWC 1/3.

[45] Gillen 1976, quoting Royal Archives Add. 9/252.

[46] Holkham MS F/TWC 10, 41.

[47] Roscoe 1833, 85.

[48] Hiskey 2005, 180.

[49] Ibid., 181, quoting Matthew Brettingham, one of the architects of Holkham.

[50] Holkham MS A54, 11.

[51] Mortlock 2007, 195 and 191, 201.

[52] Holkham MS A46.

[53] Holkham MS A48, 149.

[54] Climenson (ed.) 1899, 8.

[55] Stirling 1913, **II** 200.

[56] Ibid.

[57] MS Diary of Thomas Moore.

[58] Farington 1982, **VII** 2688.

[59] Maxwell 1903, **I** 108 and 297.

[60] Farington 1984, **XIII** 4541.

[61] Stirling 1908, **I** 238.

[62] Holkham MS A47–51, *passim*.

[63] Baring 1866, 71.

[64] Ibid., 495.

[65] Farington 1982, **VIII** 3129.

[66] Ibid., **VII** 2781.

[67] Baring 1866, 257.

[68] MS Diary of Thomas Moore.

[69] Stirling 1908, **I** 141–2.

[70] MS Diary of Thomas Moore.

[71] Stirling 1924, 263.

[72] La Rochefoucauld 1988, 193–4.

[73] Rosebery 1913, **II** 335–6.

[74] Holkham MS F/TWC 10, 70–71.

[75] Stirling 1908, **I** 242.

[76] Rosebery 1913, **II** 189.

[77] Holkham MS F/TWC 10, 52.

[78] Holkham MS Family Deeds 97.

[79] Baring 1866, 197 and 219.

[80] Farington 1979, **VI** 2311; 1982, **VII** 2545.

[81] Johnstone 1828, **VIII** 499.

[82] Lummis and Marsh 1990, 68.

[83] Sheffield Archives, SpSt 1-270 60651-2 (I am grateful to Marie-Anne Gary for drawing my attention to these documents).

[84] Gatrell 2006, 128

[85] Wraxall 1904, 531.

[86] Roscoe 1833, 81.

[87] Farington 1979, **VI** 2118.

[88] Stirling 1908, **I** 233–5.

[89] Mortlock 2007, 9.

[90] Sheffield Archives, SpSt 1-270 60651-1 and 2.

[91] Stirling 1908, **I** 234.

[92] S. Parr, *Discourse on Education*, 1786, 10 and 21.

[93] Derry 1966, xii and 35

[94] Stirling 1908, **I** 234 and 407.

[95] Stirling 1913, **II** 19.

[96] Guide book to Holkham Hall, 1817, as annotated by Dawson Turner, Thomas Phillips, B. R. Haydon and R. P. Reinagle. Trinity College, Cambridge, Library, cupboard 25, shelf 32.

[97] Climeson (ed.) 1899, 339.

[98] Illustrations in Lovell, 1995.

[99] Sheffield Archives, Sp St 1-270 60651-1.

[100] Farington 1978, **I** 144.

[101] Thorne 1986, **III** 74–5.

[102] Pickering 1903, 120.

[103] Guide to Holkham Hall, 1817, as annotated by Dawson Turner *et al.*

[104] Pickering 1903, 112.

[105] Lovell 1995, 5.

[106] Sheffield Archives, Sp St 1-270 60683.

[107] Stirling 1908, **I** 448.

[108] Trumbach 1978, 132 quoting Lord Halifax, *Complete Works*, ed. J. P. Kenyon, 1969, 288–90.

[109] Pickering 1903, 132.

[110] Delany 1861, **III** 57, quoted in Trumbach 1978, 241.

[111] Stirling 1913, **II** 51.

[112] Barney 2000, 8.

[113] Ibid., 9–10.

[114] Farington 1979, **III** 929.

[115] MS Diary of Thomas Moore.

[116] Rosebery 1913, **I** 211.

[117] Ibid., **II** 81–2.

[118] Barney 2000, 24.

[119] Farington 1979, **VI** 2118.

[120] Barney 2000, 52.

[121] Ibid., 55–6.

[122] Farington 1982, **VII** 2642.

[123] Farington 1983, **VIII** 3049.

INTERLUDE

[1] Mowl 2006, 63.

[2] *Norwich Mercury*, 8 November 1788.

[3] Holkham MS F/TWC 24, loose papers.

[4] Holkham MS F/TWC 26, loose papers.

[5] Holkham MS F/TWC 24, loose papers.

[6] Stirling 1908, **I** 347; Holkham MS F/TWC 24.

[7] Holkham MS A46, 196.

CHAPTER 4

[1] Kames, 1787, xviii.

[2] Belhaven 1699, Preface.

[3] Wade Martins 2004, 8.

[4] Young 1783, 382.

[5] An increasing amount of evidence is coming to light demonstrating the importance of 'share cropping', where the tenant and landlord shared both the costs and the profits of the crops and livestock, but by the late eighteenth century examples are generally found on smaller and more pastoral estates (Griffiths 2004).

[6] Marshall 1787, I 6.

[7] Young 1771, II 150.

[8] James 1929, 306.

[9] For a detailed analysis of the development of the Holkham estates under the first Earl of Leicester and his widow, see Parker 1975, 37–70 and Mortlock 2007, 51–60.

[10] Young 1771, I 172–5.

[11] Ibid., 177.

[12] Ibid., 178.

[13] Parker 1975, 128.

[14] Farington 1982, VIII 3050–1.

[15] W. White, *History, Gazetteer and Directory of Norfolk*, 1845, 377, Sheffield.

[16] Holkham MS E/C1/1 1816, 145.

[17] NRO WLS XXIX/6/15 416 x 4, Survey by R. Cauldwell of the Norfolk estates of Lord Walsingham, 1782.

[18] R. Gardiner, 1778 *Observations on the conduct of Thomas William Coke from his appointment of the author to be Auditor General over all his estates in Norfolk, August 1st 1776*, Holt.

[19] Ibid.

[20] Farington 1982, VIII 3050–1.

[21] NRO WKC 7/99 404 x 5; 5/396–415 440 x 7.

[22] Mary Humphrey, 1828 *A Letter to Thomas William Coke shewing the distress and misery which have been brought upon John Humphrey, Mary, his wife and their five children by the unjust treatment they have experienced from Thomas William and Lady Anne Coke*, Norwich.

[23] Holkham MS E/G20.

[24] Holkham MS E/G9.

[25] Holkham MS E/C1/3 1816, 266.

[26] Parker 1975, 90–1; Holkham MS Warham Deeds, 144c; Egmere Deeds, 5.

[27] Holkham MS E/C1/11 1824, 128–9.

[28] Parker 1975, 177–8.

[29] Ibid. 179–182; Holkham MS A/Au, 39–98.

[30] Stirling 1913, II 50.

[31] Ibid., 54.

[32] Humphrey, 1828 *A Letter to Thomas William Coke shewing the distress and misery*.

[33] Parker 1975, 43–51.

[34] Mortlock 2007, 37.

[35] Holkham MS E/G1, 16, 49, 53, 57, 64, 69.

[36] Holkham MS E/G2, 97.

[37] Holkham MS E/G1, 16, 64, 69, 83.

[38] Holkham MS E/G2, 76.

[39] Holkham MS E/G4.

[40] Holkham MS E/G5.

[41] Holkham MS E/G8, 18, 43 and 46.

[42] Bacon 1844, 290.

[43] Holkham MSS E/G1 and E/G4, *passim*.

[44] Holkham MS A/Au, 88–92.

[45] Holkham MS E/G9, 16.

[46] Holkham MS E/F1/4

[47] Young 1804, 437.

[48] Rosselli 1971, 42–64.

[49] Garrett 1749, *passim*.

[50] Holkham MS E/G20. One last equals 80 bushels or nearly 3,000 litres.

51 Kent 1775, 152.

52 Holkham MSS E/G1 and E/G4, *passim*.

53 Holkham MSS E/G1 and E/G5.

54 Holkham MS E/G7.

55 Holkham MS E/G8.

56 Young 1783, 381.

57 Wade Martins 2002, 64.

58 Young 1804, 19–20.

59 Holkham MS E/G9, 33.

60 Holkham MS E/G/14 1827, 86.

61 Holkham MS E/C1/20, 139.

62 Holkham MS A/Au, 45, 14.

63 Young 1783, 382.

64 Holkham MS E/G9, 27.

65 Holkham MS E/G11.

66 Wade Martins 1980, 132.

67 Kent 1775, 152.

68 Kent 1813, 110.

69 Wade Martins 2002, 68–111.

70 Young 1804, 32.

71 NRO WKC5/396–415 440 x 7.

72 Robinson 1979, *passim*.

73 Wade Martins 1980, 149–50.

74 Holkham MS E/G9, 15.

75 Young 1804, 20.

76 Wade Martins 2002, 52–6.

77 Ibid., 59.

78 Wade Martins and Williamson 1999, 184.

79 Parker 1975, 55.

80 Kent 1813, 223–5.

81 Parker 1975, 104.

82 Holkham MS E/G19.

83 Holkham MS E/G/8

84 NRO MS34018, 'Agricultural Journal, Memoranda, etc. as continued from my former book, J.P. Leeds'.

85 Holkham MS E/C1/3 1816, 132.

86 Ibid., 133.

87 NRO MS34018.

88 Young 1804, 398.

89 Holkham MS F/TWC 2, unnumbered pages.

90 Ibid.

91 For further discussion of water meadows in Norfolk, see Wade Martins and Williamson 1994.

92 Young 1783, 355.

93 Young *Annals of Agriculture* 39 1803, 322.

94 BL Add. MSS 35131, 10 April 1812.

CHAPTER 5

1 Young 1804, 32.

2 *Gentleman's Magazine* (1752) 22 453–5 and 502–3

3 Young 1771, II 150–63.

4 Ibid., 1–31.

5 Kerridge, 1967; Holderness 1984; Campbell and Overton 1993.

6 Wade-Martins and Williamson 1999, 5–6.

7 Holkham MS Deeds 1067.

8 Spencer 1842, 1–2.

9 Young 1783, 354.

10 Ibid., 358–9.

11 Young 1793 'A week in Norfolk' *Annals* 19 447.

12 Ibid., 457.

13 Holkham MS E/G20.

14 Holkham MS F/TWC 2.

15 Holkham MS E/G20.

16 Young *Annals* 19 1792, 457.

17 Holkham MS E/G20.

18 Young 1783, 356.

[19] Holkham MS E/F2.

[20] Young *Annals* **19**, 1793, 445.

[21] MS Dairy of Thomas Moore of Warham 1799–1811. Norfolk Family History Centre, Norwich.

[22] Tollett's MS description, privately owned, quoted by kind permission of the owner.

[23] Carter 1964, 58.

[24] Carter 1964, 305, quoting Sir Joseph Banks to Robert Fulke Greville, 2 August 1805, BM.A.MSS 42072, 71–2.

[25] Holkham MS F/TWC 2.

[26] Holkham MS E/X2.

[27] Holkham MSS Deeds 1067 and E/F/1/1.

[28] Gazley 1973, 411.

[29] Wade-Martins 1993, 25–35, 62–3.

[30] Holkham MS F/TWC 2, unnumbered pages.

[31] Young 1783, 375. I am grateful to Professor E. C. T. Collins for drawing my attention to this revival of the use of oxen.

[32] Young 1771, **I** 173.

[33] Young 1804, 481.

[34] Holkham MSS Deeds 1067, E/F1/1 and E/C1/2.

[35] Young 1804, 449.

[36] Young 1800 Duke of Bedford's Premiums for 1800, *Annals* **33**, 499–500.

[37] Betham-Edwards 1898, 385.

[38] Carter 1964, 285.

[39] Bacon 1821, 3.

[40] *Norfolk Chronicle*, 27 June 27, 1807.

[41] Rush 1873, 223.

[42] MS Dairy of Thomas Moore.

[43] Holkham MS F/TWC 2.

[44] Betham Edwards 1898, 386.

[45] Bacon 1821, 106–7.

[46] Spring 1963, 46.

[47] Holkham MS Deeds 1067.

[48] BM Add. MS 35127, f.165, 23 July 1792.

[49] Tollet's MS description.

[50] Holkham MS E/F1/4, 1817, 1.

[51] Holkham MS E/C1/1, 71.

[52] Holkham MS E/F/1/3, 43.

[53] Holkham MS E/G9, 13.

[54] Young 1783, 354.

[55] Ibid., 370.

[56] Holkham MS E/C1/1 121, Feb 9 1829.

[57] Wade Martins 2005, 135–6.

[58] *Norfolk Chronicle*, 20 and 27 November 1830.

[59] Holkham MS F/TWC 1/2, 111.

[60] Stirling, 1980, **II** 394–5; the original to which this refers cannot be located at Holkham.

[61] Holkham MS A54, 43.

[62] Holkham MS E/C1/1, 215.

[63] Holkham MS E/C1/2, 37–38.

[64] *Annals* 1793 'Substance of Sir John Sinclair's Speech to Parliament on the 1st May 1793 when he proposed the Establishment of the Board of Agriculture', **21**, 131.

[65] MERL RASE B 1.

[66] Young 1804, xv.

[67] Kent 1813, Young 1804, *passim*.

[68] MERL RASE B 7.

[69] Holkham MS F/TWC 2.

[70] *Bury and Norwich Gazette*, 18 July 1817.

CHAPTER 6

[1] Holkham MS: F/TWC 13.

[2] BL Add. MS 47580, f.189.

[3] *Parliamentary Debates* 1790, **28** 351.

[4] Hayes 1957, 238.

[5] Copeland 1967, **VI** 40.

[6] Ketton Cremer 1982, 224.

[7] Baring 1866, 197.

[8] Thorne 1986, **I**, 477.

[9] *Parliamentary Debates* 1791, **29** 37.

[10] Quoted in Hague 2004, 287.

[11] Rosebery 1913, xi.

[12] Johnstone 1828, **VII** 231.

[13] *Parliamentary Debates* 1792, **33** 92.

[14] *Norfolk Chronicle*, 1 July 1792.

[15] Johnstone 1828, **VII** 231.

[16] Ibid., 234.

[17] Ibid., 235.

[18] Russell 1853–7, **III** 89.

[19] *Parliamentary Debates* 1794, **38** 72.

[20] *Norfolk Chronicle*, 19 April 1794.

[21] *Parliamentary Debates* 1795, **41** 143.

[22] Rawcliffe and Wilson 2004, 188.

[23] *Parliamentary Debates* 1795, **43** 357.

[24] Ibid. 1797, **45** 698.

[25] Ibid., **46** 178.

[26] Wade Martins 1980, 12.

[27] *Parliamentary Debates* 1797, **46** 513.

[28] Ibid. 1800, **12** 39.

[29] Rawcliffe and Wilson 2004, 188.

[30] *Parliamentary Debates* 1795, **41** 326.

[31] Ibid., **44** 55.

[32] Ibid. 1797, **46** 656.

[33] *Norfolk Chronicle*, 21 April 1796.

[34] Thorne 1986, **II** 285.

[35] Holkham MS F/TWC/ 1/1 37.

[36] Rosebery 1913, **II** 11.

[37] Hathaway 1806 4, 208.

[38] NRO WKC/6/67.

[39] Barring 1866, 460.

[40] Holkham MS F/TWC 1/2, 69.

[41] Stirling 1908, **II** 8.

[42] Ketton Cremer 1982, 247.

[43] The only evidence for this is Stirling 1908, **II** 39.

[44] *Parliamentary Debates* 1805, **3** 363.

[45] There is no direct evidence for this, but see unsubstantiated quotations in Stirling 1908, **II** 48.

[46] Johnstone 1828, **VII** 245.

[47] Ibid., 232.

[48] Derry 1966, 251.

[49] Holkham MS F/TWC 1/1, 124.

[50] *Parliamentary Debates* 1806, **5** 523.

[51] *Norwich Mercury*, 4 October 1806.

[52] Ibid., 15 August and 8 November 1806.

[53] Ibid., 1 November 1806.

[54] Ibid., 8 November 1806.

[55] Ibid., 22 November 1806.

[56] Ibid., 15 November 1806.

[57] Stirling 1908, **II**, 76, quoting Mrs Thistlethwaite, *Memoirs and Correspondence of Henry Bathurst, Bishop of Norwich*, 1853,172.

[58] Farington 1982, **VIII** 3150.

[59] Ibid., 3152.

[60] *Parliamentary Debates* 1808, **11** 433–4.

[61] Ibid. 1809, **13** 493.

[62] GRO D678/1/F15/3/1–11.

[63] Holkham MS F/TWC 1/1, 36.

[64] Wade Martins 1980, 12.

[65] *Parliamentary Debates* 1814, **27** 891–2.

[66] Ibid., 1071.

[67] *Norfolk Chronicle*, 18 March 1815.

[68] Holkham MS F/TWC 1/2, 32.

[69] *Parliamentary Debates* 1816, **32** 55.

[70] Ibid., 1047.

[71] Ibid., 1816, **33** 153 and 155.

[72] Ibid., **34** 299 and 361.

[73] Ibid., **33** 458.

[74] Holkham MS F/TWC 1/1, 38–46.

[75] Hayes 1957, 342; *Norwich Mercury*, 11 September 1819.

[76] *Norwich Mercury*, 3 June 1826.

[77] *Parliamentary Debates* 1816, **34** 505.

[78] Ibid. 1817, **35** 759.

[79] Ibid., **35** 782.

[80] *Norfolk Chronicle*, 12 April 1817.

[81] G. Burgess, 'Mr Burgess's letter to Thomas William Coke, MP on a speech in Norwich, April 5th 1817', 1817, 22–4; G. Glover, 'An Answer to a clergyman's letter …', 1817.

[82] *Parliamentary Debates* 1818, **39** 1077.

[83] Ibid., **38** 100.

[84] Ibid., 1819, **39** 1081–2.

[85] Ibid., **39** 657.

[86] Ibid., **41** 643.

[87] Ibid., 806.

[88] O'Gorman 1997 263.

[89] *Parliamentary Debates* 1819, **41** 814.

[90] Ibid. 1822, new series **6** 96.

[91] Holkham MS F/TWC 4, unnumbered paper.

[92] Hilton 1977, 100–1.

[93] Holkham MS F/TWC 1/2, 104–5.

[94] Bodleian Library MS Eng.Hist c147, Diary of Frederick Madden, 7 March 1827.

[95] *Norwich Mercury*, 19 January 1822.

[96] *Norfolk Chronicle*, 13 April 1822.

[97] *Parliamentary Debates* 1822, new series **7** 143 and 779.

[98] *Norfolk Chronicle*, 4 June 1822.

[99] *Norwich Mercury*, 18 May 1822.

[100] *Parliamentary Debates* 1823, new series **8** 1071 and 1825, new series **13** 20.

[101] Ibid. 1829, new series **20** 992.

[102] Johnstone 1828, **VII** 270.

[103] *Norwich Mercury*, 11 January 1823, Norfolk Chronicle, 26 April 1823.

[104] *Parliamentary Debates* 1823, new series **8** 1254–7.

[105] *Norfolk Chronicle*, 4 January 1823.

[106] *Parliamentary Debates* 1823, new series **8** 1258.

[107] *Norwich Mercury*, 25 January 1823.

[108] *Statutes of the Realm* 1829 9 George IV 69.

[109] Hayes 1957, 314.

[110] *Norfolk Chronicle*, 1 October 1830.

[111] *Parliamentary Debates* 1831, third series **3** 121.

[112] *Norwich Mercury*, 11 October 1831.

[113] *Parliamentary Debates* 1832, third series **13** 302.

[114] *Norwich Mercury*, 20 April 1833.

[115] Ibid., 20 December 1832.

[116] White's *History, Gazeteer and Directory of Norfolk* Sheffield 1845, 17.

[117] Stevens 2005, 21.

[118] Holkham MS C/SFC.

[119] Maxwell 1903, **II** 294

[120] *Norwich Mercury*, 20 December 1832.

CHAPTER 7

[1] Stirling 1913, **II** 31.

[2] *Annals of Agriculture* 1793 J. Boys 'Agricultural minutes taken during a ride through the counties of Kent, Essex, Suffolk, Norfolk, Cambridge, Rutland, Leicester, Northampton, Buckingham, Bedford, Hertford, Middlesex, Berkshire and Surrey in 1792', **19** 115.

[3] Stirling 1913, **II** 59 and 43.

[4] Quoted in Schmidt *et al.* 2005, 219.

[5] Ibid., 220.

[6] La Rochefoucauld 1988, 194 and 186.

[7] Holkham MS J. Dawson *The Strangers' Guide to Holkham, Burnham, 1817*. Copy in Holkham Archives. Unfortunately the Visitors' Book does not survive.

8 Goodman 2007, 29 and 165.

9 Ibid., 73.

[10] La Rochefoucauld 1988, 196.

[11] Stirling 1913, **II** 31, 39 and 55.

[12] Holkham MS C/AN/22–3.

[13] Holkham MS F/TWC 1/3, *passim*.

[14] Bacon 1821.

[15] Holkham MS F/TWC 1/4, 34.

[16] Stirling 1908, **I** 307.

[17] Bacon 1821, 73.

[18] Rush 1873, 223 and 104–5.

[19] Holkham MS F/TWC 1/3, 9.

[20] James 1929, 292.

[21] Roscoe 1833, **II** 80–1.

[22] Roscoe 1833, **II** 57.

[23] Ibid., **II** 83.

[24] Ibid., 85–6.

[25] Ibid., 90.

[26] Holkham MS F/TWC 5.

[27] Mortlock 2006, 102.

[28] Holkham MS F/TWC 5.

[29] Roscoe 1833, **II** 264.

[30] Bodleian Library MSS Eng.hist c147, Journal of Frederick Madden, 11 March 1827.

[31] Ibid., 19 January 1827

[32] Lovell 1995, 30–1; Diary of Frederick Madden, March 1827, *passim*.

[33] La Rochefoucauld 1988, 196.

[34] Sir Alfred Shannon, 'Introduction' in Chandler 1953, xxxvi.

[35] Roscoe 1833, **II** 264.

[36] Quoted in Stirling 1908, **I** 234–5.

[37] Derry 1966, 249.

[38] Holkham MS TWC/4, loose papers.

[39] Holkham MS F/TWC 1/1.

[40] Stirling 1908, **II** 21.

[41] Holkham MS F/TWC 1.

[42] Gillen 1976, *passim*.

[43] NRO WGN 1/8.

[44] Quoted by Mortlock 2006, 98.

[45] Stirling 1908, **II** 187.

[46] Gore 1963, 181.

[47] Gillen 1976, 173.

[48] Jerningham 1896, **II** 144.

[49] Stirling 1913, **II** 128.

[50] Pickering 1903, 246.

[51] Stirling 1924, 204.

[52] Stirling 1908, **II** 186.

[53] Stirling 1913, **II** 31 and 41.

[54] Holkham MS C/AN 23.

[55] Gore 1963, 181.

[56] Pickering 1902, 390.

[57] Albemarle 1877, 209.

[58] Keppel 1899, 7–8.

CHAPTER 8

[1] Holkham MS F/TWC 1/1, 64.

[2] Pickering 1902, 102.

[3] Pickering 1902, 127.

[4] Stirling 1908, **II** 283.

[5] Seymour 1906, 244.

[6] Stirling 1913, **II** 17.

[7] Stirling 1908, **II** 283.

[8] *Parliamentary Debates* 1822, 6 986.

[9] Stirling 1913, II 39.

[10] Holkham MS C/AN 20.

[11] Stirling 1908, II fn 306.

[12] Stirling 1913, II 11.

[13] Pickering 1903, 136 and 141.

[14] Stirling 1913, II 31.

[15] Pickering 1902, 142–3.

[16] Stirling 1913, II 43 nad 55.

[17] Stirling 1913, II 48, 53, 54 and 55.

[18] Holkham MS A52, 91.

[19] Holkham MS
A54 1831–1834, 15, 63, 107, 156.
A55 1835–1838, 16, 60, 107, 158.
A56 1839–1842, 17, 61, 96, 135.

[20] Mortlock 2006, 111.

[21] Holkham MS F/TWC 1/1, 35.

[22] Maxwell 1903, II 111.

[23] Gore 1963, 436.

[24] Coke 1905, 77.

[25] Stirling 1924, 204.

[26] Stirling 1913, II 74–6, 82, 105.

[27] Mary Humphrey, 'A Letter to Thomas William Coke shewing the distress and misery which have been brought upon John Humphrey, Mary, his wife and their five children by the unjust treatment they have experienced from Thomas William and Lady Anne Coke', Norwich, 1828.

[28] Maxwell 1903, II 331.

[29] Pickering 1902, 154.

[30] Gore 1963, 438

[31] Pickering 1902, 148.

[32] Ibid., 151.

[33] Holkham MS EC/9,126–7.

[34] Reproduced as Appendix 4 in Parker 1975.

[35] Spring 1963, 32–3, Becket 1994, 178, Table 7.

[36] *Norfolk Chronicle*, 29 January 1820.

[37] Ibid., 27 January 1821.

[38] Ibid., 26 January 1822.

[39] Le Strange 1896, 272.

[40] Stirling 1908, I 195–6.

[41] Le Strange 1896, 324.

[42] Ilchester 1937, 175.

[43] Holkham MS: C/AN 16.

[44] Holkam MS C/AN 13.

[45] Holkham MS C/AN 12–13.

[46] Coke 1905, 67.

[47] Holkham MS C/AN 17.

[48] Miller 2005, 151.

[49] Coke 1905, 14.

[50] Holkham MS C/AN 17.

[51] Holkham MS A54, 22.

[52] BL Add. MS 51593, Holland Papers.

[53] Holkham MS C/AN 14.

[54] Holkham MS C/AN 21.

[55] Holkham MS C/AN 26.

[56] Holkham MS C/AN 19.

[57] Ilchester 1937, 176.

[58] Ibid., 175.

[59] Pickering 1903, 246.

[60] *Norwich Mercury*, 15 January 1841.

EPILOGUE

[1] *Norfolk Chronicle*, 21 August 1847

[2] Armstrong 1963, 63; Wade Martins 2005, 582–3.

[3] Bacon 1845, *passim*.

[4] Stirling 1913, I 216.

[5] Ibid.

[6] *Norfolk Mercury*, 16 August 1845.

BIBLIOGRAPHY

A Note on Sources

There has been no biography of Thomas William Coke to supersede the two-volume work by his great-granddaughter, Mrs Stirling, published in 1908. She had full use of the archives and may indeed have borrowed some material that is now missing.

While the estate archives at Holkham are remarkably complete, there are far fewer personal papers. These include two MS incomplete memoirs, which may have been dictated by Coke himself in his old age, details of electioneering and of the arrangements for the 1788 celebrations, collections of letters and miscellaneous papers. The fact remains, however, that much of our information is derived from Coke and what he chose to recall for his two contemporary biographers.

Primary Sources at Holkham

A46–56 Household Accounts, 1782–1842.

A/Au 38–104 Audit Books, 1776–1842.

C/AN Papers of Alexander Napier.

C/CFS Frances Chantrey letters.

E/C1/1–2 Agricultural Letter books.

E/C1/3–30 Letter books, 1816–1842.

E/F1–4 Home Farm Accounts, 1814–1826.

E/G1 Dugmore's Plans and Particulars of the Norfolk Estates, 1778.

E/G2 Wyatt's Survey of T. W. Coke's Estates, 1779.

E/G4 Biederman's Plans and Particulars of the Norfolk Estates, 1781.

E/G5 Biederman's Plans Revised by Dagmore, 1789.

E/G7 Nathaniel Kent's Survey of Warham, 1785–1788.

E/G8 Field Book 1789–1802.

E/G9 Blaikie's Survey of the Estate 1816.

E/G11 Leak's Plans and Particulars of the Estates Belonging to Thomas William Coke, 1828.

E/G20 Book of Observations, 1801–1858.

E/W Sandys's Account of Trees, c.1781–1804.

E/X2 Sheepshearing Account, 1815–16.

E/X3 Sheepshearing Account, 1804–7.

E/X4 Sheepshearing Account, 1804–1813.

F/TWC 1/1–4 Bound Volumes of Letters, c.1776–1837.

F/TWC 2 Loose Bundles of Letters, c.1776–1840.

F/TWC 4 Samuel Parr's Letters to Thomas William Coke, 1790–1829.

F/TWC 5 Letters of William Roscoe, 1812–1833.

F/TWC 10 'Memoir of Thos. Wm. Coke, MP up to 1783', writer unknown (possibly R. N. Bacon), n.d.

F/TWC 11 T. W. Coke by the Revd Thomas Keppel, n.d.

F/TWC 12 Correspondence with Sir Francis Chantrey, 1830–1835.

F/TWC 13 Humphrey Repton 'A General View of the Influences Operating on Elections in the County of Norfolk, n.d. c.1790.

F/TWC 14–18 Correspondence, etc. Concerning Elections, 1776–1796.

F/TWC 19–23 Correspondence, etc. Concerning Elections, 1784 and 1796.

F/TWC 24–29 Correspondence, etc. Concerning the Glorious Revolution Centenary Ball, 1788.

Repton's 'Red Book' 1789 (Holkham Archives).

Family Deeds 1776–1842 (Holkham MS).

Game Book 1793–1842 (Holkham Hall).

Primary Sources held elsewhere

Bodleian Library, Oxford
Journal of Frederick Madden MSS Eng hist c.140–182 1814–1872.

British Library
BM Add MS 35127 Correspondence of Arthur Young.

Gloucester Record Office (GRO)
D678/1E1/3/ Rental of the estates of the Dutton family of Sherborne and Standish 1732–1776.

1F15/3/1–11 letters relating to the election of John Dutton.

1F7/1/9–12 Marriage Settlements 1769–1822.

Museum of English Rural Life (MERL)
RASE B Minute books of the Board of Agriculture, vols 1 to 8 1797–1822.

Norfolk Family History Centre, Norwich
Transcript of diary of Thomas Moore Husdon of Warham, Norfolk 1799–1811.

Norfolk Record Office (NRO)
WGN 2/1 Diary of the Revd William Gunn 1792–1793.

    3/1 Journal of Ann Gunn, 1st May to 29th November 1792.

WKC 6/67 401x7 Letter to William Windham III from Thomas Amyot.

WLS XXIX/6/15 416x4 Survey by Ralph Cauldwell of the Norfolk Estates of Lord Walsingham 1782.

MS 34018 'Agricultural Journal, Memoranda etc. as continued from my former book by J. P. Leeds' 1823–1828.

Sheffield City Archives
SpSt 1-270 60651-1 Spencer Stanhope Munuments, Jane Coke's Account books 1789–1800.

Printed Sources

Albemarle, Earl of (1877 edn) *Fifty Years of My Life*, 3 vols. London, Macmillan.

Armstrong, H. B. J. (ed.) (1963) *Armstrong's Norfolk Diary*. London, Hodder and Stoughton.

Bacon, R. N. (1821) *Report of the Transactions at the Holkham Sheep Shearings*. Norwich.

Bacon, R. N. (1845) 'Narrative of the Proceedings Regarding the Erection of the Leicester Monument Norwich', Norwich.

Bacon, R. N. (1844) *The Agriculture of Norfolk*. London, Ridgeways, Piccadilly, and Chapman and Hall, Strand.

Baring, Mrs H. (1866) *The Diary of the Right Hon. William Windham 1784–1810*. London, Longmans.

Barney, J. (2000) *The Defence of Norfolk 1793–1815*. Norwich, Mintaken.

Barringer, J. C. (ed.) (1989 reprint) *Fadens Map of Norfolk*. Stibbard, Larks Press.

Beckett, J. V. (1994) *The Rise and Fall of the Grenvilles*. Manchester University Press.

Betham-Edwards, M. (ed.) (1898) *Autobiography of Arthur Young*. London, Smith, Elder & Co.

Belhaven, Lord (1699) *The Countreymans Rudiments, or An Advice to the Farmers of East Lothian on How to Labour and Improve Their Ground*. Edinburgh, A. Anderson.

Black, J. (1999) *The British Abroad: The Grand Tour in the Eighteenth Century*. London, Sandpiper Books.

Brander, M. (1964) *The Hunting Instinct: The Development of Field Sports over the Ages*. London, Oliver & Boyd.

Brooke, J. (1968) *The History of Parliament: The Commons 1754–1790*. Vol. I, *Introductory Survey*. Oxford University Press.

Campbell, B. M. S. and Overton, M. (eds.) (1991) *Land, Labour & Livestock, Historical Studies in European Agricultural Productivity*, Manchester University Press.

Card, T. (2001) *Eton Established*. London, John Murray.

Carter, H. B. (1964) *His Majesty's Spanish Flock*. London, Angus and Robertson.

Chandler, G. (1953) *William Roscoe of Liverpool*. London, Batsford.

Christie, Ian R. (1957) 'Thomas William Coke and American Independence', *Norfolk Archaeology* 31, 417–18.

Christie, I. R. (1982) *Wars and Revolutions, Britain 1760–1815*. London, John Arnold.

Climenson, E. J. (ed.) (1899) *Passages from the Diaries of Mrs Lybbe Powys of Harwick House, Oxfordshire AD 1756–1808*. London, Longmans.

Coke, Henry (1905) *Tracks of a Rolling Stone*. London, Smith Elder and Co.

Coke, Lady Mary (1896) *Journal of Lady Mary Coke*, 4 vols. Edinburgh, David Douglas.

Cole, G. D. H and Cole, M. (eds) (1930) William Cobbett's *Rural Rides*, 3 vols. London, P. Davies.

Colley, L. (2003) *Britons. Forging the Nation 1707–1837*. London, Pimlico.

Copeland, T. (ed.) (1967) *The Correspondence of Edmund Burke*, vol. VI. Cambridge University Press.

Cunningham, H. (1989) 'The Language of Patriotism' in *Patriotism: The Making and Unmaking of National Identity*, ed. R. Samuel. London, Routledge.

Curwen, J. C. (1809*) General Hints on Agricultural Subjects*. Reprinted Cambridge, Chadwyck-Healey, 1990.

Cust, L. (1899) *A History of Eton College*. London, Duckworth.

Defoe, Daniel (1971) *A Tour through the Whole Island of Great Britain*. London, Penguin Books.

Delany, M. (1861) *Autobiography and Correspondence of Mary Granville, Mrs Delany*. Lady Llanover (ed.) 3 vols. London, Richard Bentley.

Derry, W. (1966) *Dr Parr: A Portrait of a Whig Dr Johnson*. Oxford, Clarendon Press.

Ehrman, J. (1969) *The Younger Pitt*, 3 vols. London, Constable.

Farington, J. (1978–84) *The Farington Diaries*, vols I–VI, ed. K. Garlick and A. MacIntyre; vols VII–XV, ed. K. Cave. London and New Haven, Yale University Press.

Garrett, D. (1749) *Designs for Farm Houses etc. for the Counties of Yorkshire, Northumberland, Cumberland, Westmorland and the Bishopric of Durham*.

Gatrell, V. (2006) *City of Laughter*. London, Atlantic Books.

Gazley, J. G. (1973) *The Life of Arthur Young*. Philadelphia, American Philosophical Society.

Gillen, M. (1976) *Royal Duke*. London, Sidgwick & Jackson.

Gore, J. (ed.) (1963) *The Creevey Papers*. London, Batsford.

Girouard, M. (1978) *Life in the English Country House*. New Haven and London, Yale University Press.

Goodman, N. (ed.) (2007) *Dawson Turner: A Norfolk Antiquary and His Remarkable Family*. Chichester, Phillimore.

Gore, A. and Carter, G. (eds) (2005) *Humphrey Repton's Memoirs*. Norwich, Michael Russell Publishing Ltd.

Griffiths, D. (2002) *Patriotism and Poetry in Eighteenth Century Britain*. Cambridge University Press.

Griffiths, E. (2004) 'Responses to Adversity: The Changing Strategies of Two Norfolk Landowning Families, *c.* 1665–1700', in *People, Landscape and Alternative Agriculture. Essays for Joan Thirsk*, ed. R. W. Hoyle. Agricultural History Society Supplement Series, 5.

Hague, W. (2004) *William Pitt*. London, HarperCollins.

Halfpenny, W. (1750) *Twelve Beautiful Designs for Farm Houses*. London.

Harling, P. (1996) *The Waning of 'Old Corruption'. The Politics of Economical Reform in Britain, 1779–1846*. Oxford University Press.

Hathaway, W. S. (1806) *The Speeches of the Right Honourable William Pitt in the House of Commons*, 4 vols. London, Longman.

Hayes, B. D. (1957) 'Politics in Norfolk 1750–1832', University of Cambridge DPhil dissertation.

Hilton, B. (1977) *Corn, Cash, Commerce: The Economic Policies of Tory Governments 1815–1830*. Oxford University Press.

Hiskey, C. (2005) 'Living at Holkham', in *Holkham*, ed. L. Schmidt, C. Keller and P. Feversham. London, Prestel, 176–86.

Holderness, B. A. (1984) 'East Anglia and the Fens' in J. Thirsk (ed.) *Agrarian History of England and Wales*, Vi 197–238. Cambridge University Press.

Ilchester, Earl of (1937) *Chronicles of Holland House 1820–1900*. London, John Murray.

James, C. W. (1929) *Chief Justice Coke, His Family and Descendants at Holkham*. London, Country Life.

Jerningham, F. (1896) *The Jerningham Letters 1780–1843*, ed. E. Castle. London, R. Bentley and Son.

Jewson, C. B. (1975) *Jacobin City*. Glasgow, Blackie.

Johnson, C. W. (1843) 'Memoirs of the Late Earl of Leicester', *Farmers' Magazine*, 1–6.

Johnstone, J. (1828) *The Works of Samuel Parr LLD with Memoirs of His Life and Writings*, 8 vols. London, Longman.

Kames, Lord H. H. (1787) *The Gentleman Farmer*. Edinburgh, Creech.

Kelly, Paul (1972) 'Radicalism and Public Opinion in the General Election of 1784', *Bulletin of the Institute of Historical Research* 45, 72–88.

Kent, N. (1775) *Hints to Gentlemen of Landed Property*. London, J. Dodsley, Pall-Mall.

Kent, N. (1794) *A General Survey of the Agriculture of the County of Norfolk*. London, C. Macrae.

Kent, N. (1813) *A General Survey of the Agriculture of the County of Norfolk*, second edn. London, Sherwood, Neely and Jones.

Keppel, Sir H. (1899) *A Sailor's Life under Four Sovereigns*. London, Macmillan.

Kerridge, E. (1967) *The Agricultural Revolution*. London, Allen and Unwin.

Ketton Cremer, R. W (1982) *Felbrigg, the Story of a House*. London, Futura.

Kingley, N. (1992) *The Country Houses of Gloucestershire*, vol. II, *1500–1660*. Chichester, Phillimore.

La Rochefoucauld, François, comte de (1988) *A Frenchman's Year in Suffolk 1784*, ed. and trans. N. Scarfe. Suffolk Record Society 30. Woodbridge, Boydell.

Le Strange, H. (1896) *A History of Freemasonry in Norfolk*. Norwich, A. H. Goose.

Lovell, Mary S. (1995) *A Scandalous Life: The Biography of Jane Digby el Mezral*. London, Richard Cohen Books.

Lummis, T. and Marsh, J. (1990) *The Woman's Domain: Women and the English Country House*. London, Viking.

Macnaughton, D. A. (1996) *Roscoe of Liverpool: His Life, Writings and Treasures*. Birkenhead, Countyvise.

Marshall, William (1787) *The Rural Economy of Norfolk*, 2 vols. London, Cadell.

Maxwell (ed.) (1903) *The Creevey Papers*, 2 vols. London, Murray.

Miller, G. F. (2005) 'Edward Ellice', *New Dictionary of National Biography*. Oxford University Press, 18, 151.

Mingay, G. E. (1963) *English Landed Society in the Eighteenth Century*. London, Routledge and Kegan Paul.

Mitchell, L. G. (1992) *Charles James Fox*. Oxford University Press.

Mitchell, L. G. (2005) 'Charles James Fox', *New Dictionary of National Biography*. Oxford University Press, 20, 619–22.

Moore, A. (1985) *Norfolk and the Grand Tour*. Norwich, Norfolk Museums Service.

Mortlock, D. P. (2006) *Holkham Library: A History and Description*. London, Roxborough Club.

Mortlock, D. P. (2007) *Aristocratic Splendour: Money and the World of Thomas Coke, Earl of Leicester*. Stroud, Sutton Publishing.

Mowl, T. (2006) *William Kent: Architect, Designer, Opportunist*. London and New Haven, Yale University Press.

Namier, L. and Brooke, J. (1964) *The History Parliament: The Commons 1754–1790*, 3 vols. London, Secker and Warburg.

O'Gorman, F. (1997) *The Long Eighteenth Century*. London, Arnold.

Overton, M. (1996) *The Agricultural Revolution in England*. Cambridge University Press.

Parker, R. A. C. (1975) *Coke of Norfolk: A Financial and Agricultural Study*. Oxford, Clarendon Press.

Pevsner, N. and Williamson, E. (1978) *Buildings of England, Derbyshire*, second edn. London, Penguin Books.

Pevsner, N. and Wilson, B. (1999) *Buildings of England, Norfolk 2: North-west and South*, second edn. London, Penguin Books.

Pickering, Spender (ed.) (1902) *Memoirs of Anna Maria Wilhelmina Pickering*. London.

Rawcliffe, C. and Wilson, R. (eds) (2004) *Norwich since 1550*. London, Hambledon.

Riviere, M. (1965) 'A Norfolk Parson on the Grand Tour', *Norfolk Archaeology* 33, 351–402.

Robinson, J. M. (1979) *The Wyatts, an Architectural Dynasty*. Oxford University Press.

Robinson, J. M. (1983) *Georgian Model Farms*. Oxford, Clarendon Press.

Robinson, J. M. (1988) *The English Country Estate*. London, Century.

Robinson, J. M. (1991) *A Guide to the Country Houses of the North West*. London, Constable.

Rosebery, Right Hon. the Earl of (1913) *The Windham Papers*. London, Herbert Jenkins.

Roscoe, H. (1833) *Life of William Roscoe*. London, Cadell.

Rosselli, J. (1971) 'An Indian Governor in the Norfolk Marshland', *Agricultural History Review* 19, 42–64.

Rubenstein, J. (2004) 'Lady Mary Coke', *New Dictionary of National Biography*. Oxford University Press, **12**, 468–470.

Rudge, T. (1813) *General View of the Agriculture of the County of Gloucestershire.* London, Sherwood, Neely and Jones.

Rush, R. (1873) *The Court of London 1819–1825*, third edn. London.

Russell, Lord J. (ed.) (1853–7) *Memorials of Charles J. Fox*, 4 vols. London, Richard Bentley.

Schmidt, L., Keller, C. and Feversham, P. (eds.) (2005) *Holkham*, London, Prestel.

Seymour, Lady E. (1906) *The Pope of Holland House.* London, T. Fisher Unwin.

Spencer, Earl (1842) 'On the Improvements which have taken place in West Norfolk' *Journal of the Royal Agricultural Society of England* 3, 1–9.

Spring, D. (1963) *The English landed Estate in the nineteeenth century*, Baltimore, Baltimore University Press.

Stevens, T. (2005) 'Sir Francis Chantrey', *New Dictionary of National Biography.* Oxford University Press, **11**, 21.

Stirling, A. M. W. (1908) *Coke of Norfolk and His Friends*, 2 vols. London, John Lane, The Bodley Head.

Stirling, A. M. W. (1913) *The Letter Bag of Lady Elizabeth Spencer Stanhope.* London, John Lane.

Stirling, A. M. W. (1924) *Life's Little Day.* London, Thornton Butterworth.

Stirling, A. M. W. (1960) *A Scrap Heap of Memories.* London, Macmillan.

Thompson, F. M. L. (1963) *English Landed Society in the Nineteenth Century.* London, Routledge Kegan Paul.

Thorne, R. G. (1986) *The History of Parliament: The Commons 1790–1820*, 5 vols. London, Secker and Warburg.

Trumbach, R. (1978) *The Rise of the Egalitarian Family.* New York, Academic Press.

Wade-Martins, P. (1993) *Black Faces: A History of East Anglian Sheep Breeds.* Norwich, Norfolk Museum Service.

Wade Martins, S. (1980) *A Great Estate at Work: The Holkham Estate and its Inhabitants in the Nineteenth Century.* Cambridge University Press.

Wade Martins, S. (2002) *The English Model Farm.* Macclesfield, Windgather.

Wade Martins, S. (2004) *Farmers, Landlords and Landscapes.* Macclesfield, Windgather.

Wade Martins, S. (2004) 'John Hudson' *New Dictionary of National Biography.* Oxford University Press, **28**, 582–3.

Wade Martins, S. (2005) 'Voting in the Late Eighteenth Century', 126–7; 'The Riots of 1830', 135–6, in *An Historical Atlas of Norfolk*, ed. T. Aswin and A. Davison. Chichester, Phillimore.

Wade Martins, S. and Williamson, T. (1994) 'Floated Water Meadows in Norfolk: A Misplaced Innovation', *Agricultural History Review* **42**, 20–37.

Wade Martins, S. and Williamson, T. (1999) *Roots of Change*. Agricultural History Society Supplement Series, **4**.

Walpole, H. (1937) Correspondence. W. S Lewis (ed.). London & New York Yale.

Waterson, M. (1980) *The Servant's Hall: A Domestic History of Erdigg*. London, National Trust.

Williamson, T. (1995) *Polite Landscapes: Gardens and Society in Eighteenth Century England*. Stroud, Sutton Publishing.

Williamson, T. (2005) 'The Development of Holkham Park', in *Holkham*, ed. L. Schmidt, C. Keller and P. Feversham. London, Prestel, 58–63.

Winkley, G. (1986) *The Country Houses of Norfolk*. Lowestoft, Tyndale Press/Panda Publishing.

Wraxall, Sir N. W. (1904) *Historical Memoirs of My Own Time*. London, Kegan, Paul, Trench & Trubner.

Wright, J. (1815) *Speeches of the Rt. Hon. Charles James Fox in the House of Commons*, 6 vols. London, Longmans.

Young, A. (1768) *Six Weeks Tour in the Southern Counties*. London, Nicoll.

Young, A. (1771) *The Farmer's Tour through the East of England*, 4 vols. London.

Young, A. (1783) 'A Minute of the Husbandry of Thomas William Coke', *Annals of Agriculture* **2**, 352–83.

Young, A. (1804) *General View of the Agriculture of the County of Norfolk*. London, P. and W. Nicoll, Pall-Mall.

INDEX

Printed and bound by CPI Group (UK) Ltd, Croydon, CR0 4YY

30/08/2023

08106883-0001